Coping with the Climate Crisis

COPING
WITH THE
CLIMATE
CRISIS

*Mitigation Policies and
Global Coordination*

EDITED BY

Rabah Arezki, Patrick Bolton,
Karim El Aynaoui, and Maurice Obstfeld

Columbia University Press
New York

COLUMBIA | SIPA
Center on Global Economic Governance

Columbia University Press
Publishers Since 1893
New York   Chichester, West Sussex
cup.columbia.edu
Copyright © 2018 Columbia University Press
All rights reserved

Library of Congress Cataloging-in-Publication Data
Names: Arezki, Rabah, editor. | Bolton, Patrick, 1957– editor. |
El Aynaoui, Karim, editor.
Title: Coping with the climate crisis : mitigation policies and global
coordination / edited by Rabah Arezki, Patrick Bolton,
Karim El Aynaoui, and Maurice Obstfeld.
Description: New York : Columbia University Press, [2018] |
Includes bibliographical references and index.
Identifiers: LCCN 2017049258 (print) | LCCN 2018011489 (ebook) |
ISBN 9780231547352 (electronic) | ISBN 9780231187565
(cloth : alk. paper)
Subjects: LCSH: Climate change mitigation—Economic aspects.
Classification: LCC QC903 (ebook) | LCC QC903 .C675 2018 (print) |
DDC 363.738/747—dc23
LC record available at https://lccn.loc.gov/2017049258

Columbia University Press books are printed on permanent
and durable acid-free paper.
Printed in the United States of America

Cover design: Milenda Nan Ok Lee
Cover art: Design Pics Inc/© Alamy

Contents

Foreword

WE ARE LIVING today through a period of great changes toward a new world order, triggered by a profound technological shift that impacts the very way we live and transforms radically our economic models as much as our mobilization and communication capabilities.

A new world, in which augmented reality and universal knowledge are within everyone's reach through a simple smartphone; in which virtual self-powered intermediation increasingly controls traditional production and intermediation systems; a world in which social demands are determined less and less by local, national, or regional potential and more and more in relation to the more developed prevailing lifestyles that are continuously displayed on social media and through mainstream networks.

This reconfiguration of the world order brings with it legitimate and increasingly pressing aspirations within the populations of developing countries in the south to close the gap, and have access to the means for progress and well-being as it is perceived in northern developed countries.

In this context of a massive march toward human well-being the question is posed whether our planet is capable of meeting all these needs while using the current modes of production and consumption.

As we are all aware, addressing this particular concern has triggered a global collaborative process aimed at the fight of climate change effects, which are direct results of that same pursuit of human development and well-being, while promoting new models for resilient progress with less alarming consequences and less irreversible effects on natural ecosystems.

In that context, the COP21 (the twenty-first session of the UN Conference of the Parties on Climate Change) held in Paris, marked a decisive moment with a radical shift of paradigm toward climate resilience and combating climate change effects.

It was indeed a shift of paradigm because, for the first time, all countries around the world agreed and committed to act individually and collectively to limit global warming to well below 2°C, i.e., above preindustrial levels; with each country submitting their respective voluntary contributions to achieve the objective set and committing to evolve, over time, toward a carbon-free society, in particular through energy transition.

A shift of paradigm because by acknowledging the principle of common but differentiated responsibilities of parties, a principle that is henceforth embodied in the Paris Agreement, the global community opted for an approach that makes it possible to manage and accommodate the differences among developed, developing, and least developed countries, taking into account the differentiated levels of vulnerability and contribution to climate change.

A shift of paradigm because the parties recognized the decisive role that all stakeholders must play to achieve the targets set, including subnational governments, businesses, civil society, trade unions, professional organizations, financial institutions, and citizens.

However, such a radical shift of paradigm can only become effective through actual concrete implementation. This is why the Moroccan Presidency is determined to make the COP22 the COP of concretization and action. That is our primary objective for this COP22, and there lies our main challenge.

In his Royal Message to the Medcop held in July 17, 2016 in Tangiers, His Majesty the King Mohammed VI drew up the priority areas of focus for the COP22 Moroccan Presidency:

[. . .] the priorities of the Moroccan Presidency of COP22 focus on four areas: the actual implementation of national contributions; the mobilization of funds; the strengthening of adaptation measures and technological development.

Firstly, we need to encourage countries to adopt voluntary national contributions and to translate them into integrated public policies.

Secondly, Morocco will seek to establish a process for the gradual mobilization of funding for the benefit of developing countries.

At the same time, it will propose mechanisms to facilitate access to climate finance and maximize benefits.

Thirdly, a significant effort will be made in the area of adaptation, through the quantification of needs, an increase in resources and greater capacity building.

Finally, COP22 will consider a crucial action plan devoted to technology. It will comprise three main components: the dissemination of mature technology, the emergence of transition technology and support for innovation through research and development.

As for Morocco's voluntary contributions, which were submitted for the **opinion of the Economic, Social and Environmental Council (CESE) of the Kingdom of Morocco**, which I have the honor of chairing and brings together employer representatives, the translation of such into integrated public policies trade unions, NGOs, experts, and national institutions, they can be articulated around some key areas of intervention.

First, it is essential to accelerate the **strengthening of the institutional governance of climate policy**, through effective implementation of the national integrated sustainable development strategy, which was developed based on a participative approach and rooted around the different territories of the Kingdom, and includes adequate regulations, capacity building targets with appropriate training programs for policy makers and other key actors, as well as adequate planning in terms of identification, prioritization, and budgetization of the required actions.

In that regard, it is important to recall that **policy timing** and **climatic timing** do not necessarily coincide. Indeed, although we have a general understanding of the climate change paths, the latter also include the increasingly frequent occurrence of uncontrollable extreme events. Social and political pressures related to such extreme events will inevitably lead policy makers, when faced with emergency situations, to opt for short-term solutions to the detriment of the need to safeguard the interests of future generations.

Hence, it is essential to (i) refocus governments' efforts on adaptation in their fight against climate change, (ii) align and upstream time horizons of the various sectoral policies with the long-term horizon of climate policies aimed to fight against the effects of climate change, and (iii) put in place adequate mechanisms for resource revenue allocation to the various policies.

Second, it is important **to seize** what some call **the "climate chance,"** the new economic opportunities created by the fight against climate change effects. Indeed, the various sectors of the green and blue economy offer genuine opportunities in terms of growth and jobs whether in energy transition, the circular economy, infrastructure, or associated services. One of the challenges to be addressed in this area is to enable climate-smart project holders and people with innovative ideas by facilitating access to climate financing resources; climate financing that must be more clearly identified, mobilized, and restructured with an increased focus on adaptation.

Third, endeavors to combat the effects of climate change can only succeed through efforts to **strengthen the resilience of territories and cities** in dealing with these effects. This means that the challenges of climate change must be taken on board in all land-use policies and urban planning strategies.

This involves, in particular, **densification of the urban fabric** reducing the hold of cities on arable land, and reducing the dependence of cities close to the shoreline on the sea. In parallel, and so as to be able to cope with extreme weather events, reliable crisis and natural disaster management arrangements as well as information, inventory, and sectoral and territorial systems for measuring greenhouse gases (GHGs) must be put in place.

Finally, the **fourth** area consists in acknowledging that combating climate change is not a matter that concerns only governments or experts. It requires the **mobilization and involvement of all civil society actors, whose action needs to be empowered and strengthened.** Therefore, it is crucial to institutionalize the participation of civil society at all levels from the design stage and implementation planning processes, to the monitoring and assessment of local, national, regional, and global climate policies. It is particularly important to value and capitalize on cultural heritage and on traditional and ancestral practices that are a wellspring from which we can draw inspiration, ideas, and solutions in the fight against climate change.

The timing of COP22 awakened us to the need to base our work around a **triple pact:**

- an *environmental pact*, key principles of which are enshrined in the Paris Agreement that we have committed to implement;

- a *social pact*, with at its core the fundamental principle of inter-generational solidarity at the local, national, and regional levels and between countries from the North and the South;
- and, an *innovative economic and financial pact* to smooth the transition toward a resilient and sustainable development model.

It is in the context of this triple pact that we are called upon to propose innovative, operational mechanisms to strengthen the ambitions with regard to energy transition, and improve and normalize national contributions.

The deliberations of this gathering, which focused on discussing important and challenging issues such as how to manage uncertainty, mobilization of finance, and resilient infrastructure in view of the implementation of the Paris Agreement and other relevant emerging global trends, certainly contributed to enrich our common reflection, making the Marrakech COP22 the COP of action, the COP of effective cooperation, the COP of solidarity, the COP of synergies and complementarity among all the forces of the planet to safeguard the interests of future generations.

Nizar Baraka
President of the Scientific Committee of COP22

Acknowledgments

MANY PEOPLE CONTRIBUTED to the organization of the Marrakesh conference of September 8–9, 2016, "The Energy Transition, NDCs, and the Post-COP21," out of which this book arose, and many people contributed to the production of this book.

At Columbia University we are grateful to the Center on Global Economic Governance, and especially Jan Svejnar, David Caughlin, and Theresa Murphy for all their help with the preparation of the conference.

At Columbia University Press, we are especially grateful to Bridget Flannery-McCoy, Christian Winting, and Leslie Kriesel. And to Ben Kolstad, of Cenveo, who managed the project.

At the International Monetary Fund we are grateful to Christian Bogmans, Lucia Buono, and Rahel Kidane for their help with the conference and book manuscript.

At OCP Policy Center, who hosted the conference, we are grateful to Karim El Aynaoui, Rim Berahab, and all the staff involved in the organization.

At Amundi and the Portfolio Decarbonization Coalition, we thank Timothee Jaulin for his excellent insights and help in preparing the conference.

Finally, we wish to thank the OCP Policy Center, the International Monetary Fund, and the Center on Global Economic Governance for their ongoing support of this project.

Coping with the Climate Crisis

Introduction

RABAH AREZKI, PATRICK BOLTON,
KARIM EL AYNAOUI, AND MAURICE OBSTFELD

CLIMATE CHANGE IS upon us. The globe is warming inexorably. Average temperatures set new record highs nearly every year that goes by. Sea levels are rising. Polar ice caps and glaciers are melting away. Droughts are more intense and persistent. So are wildfires. Coral reefs are bleaching, and more and more species are facing extinction. Climate change is already threatening far too many species and will become an existential threat for humanity if it is not reversed soon. Scientists have long established that the main cause of global warming is the emission of carbon dioxide particles into the atmosphere that trap heat and create a global greenhouse effect. The solution to climate change is clear: stop carbon emissions. Unfortunately, the global collective action problem of bringing humanity to stop emitting carbon is the biggest and most complex political and economic problem the world has ever faced.

It was not for want of trying, but the United Nations–led efforts initiated in 1992 through the Framework Convention on Climate Change (UNFCCC) did not make significant headway until the historic agreement reached in Paris in 2015 at the United Nations

Conference of the Parties (COP21). By now 166 parties have ratified the agreement, and these signatories have started to implement the commitments to emission reductions they agreed to in Paris. The fact that it took the world over twenty years to hammer out an agreement, and one that many commentators say falls short of what is needed to reverse global warming, testifies to the political and economic complexity of accomplishing any meaningful global collective action on carbon emissions. But the failure of the Kyoto Protocol adopted in 1997, and subsequent efforts to revive it, is not just a reflection of the difficulties of getting countries as diverse as India, Nigeria, the United States, Venezuela, Canada, Norway, Spain, and the Maldives to overcome their opposing interests to contain global emissions. The Kyoto process also failed because of the top-down approach taken to limit emissions. By the time the negotiations around the Paris climate agreement of 2015 had started, the coalition of nations leading climate change mitigation efforts had shrunk to a few countries.

The success of COP21 is largely due to a complete change of tactic. Instead of a top-down approach, in which economic agents respond to regulations imposed on them, the Paris Agreement is built on a bottom-up approach that includes all major economic actors, not just national governments. Also, the emphasis is more on inclusiveness than on immediately achieving major emission reductions, recognizing that there is not much point in getting only a small number of nations to implement an ambitious climate agreement. Finally, the approach taken to the implementation of the Paris Agreement is one of community rather than legal enforcement. The hope, which so far seems to hold, is that leadership by example, moral suasion, public opinion, influence activities by civil society and economic actors, will prod national governments, municipalities, investors, and corporations to accelerate the energy transition away from fossil fuels. This bottom-up approach is inevitably less coherent, but it is more realistic and leverages the power of civil society and business.

The Paris Agreement aims at limiting the global temperature increase this century to below 2°C. Countries and other organizations have volunteered their own contributions to limiting emissions between 2020 and 2030, the so-called "nationally determined contributions" (NDCs). The 2016 United Nations Climate Change Conference of the Parties (COP22) in Marrakesh focused on implementation issues raised by the Paris Agreement and how to deliver on the promises made in Paris. Of special importance was the issue of linking climate change mitigation steps to the United Nations Sustainable Development Goals (SDGs), recognizing that global warming also has dramatic social consequences disproportionately affecting the poor, with floods and droughts resulting in famines and the forced movement of populations. In that regard, the Kingdom of Morocco, the host of the COP22, has taken important initiatives related to sustainable agriculture, food security, renewable energy, and land use to support developing countries' efforts to guard against the impacts of climate change.

This volume is based on the proceedings of a conference held in Marrakesh, Morocco, on September 8 and 9, 2016 in the run-up to the COP22. The conference was organized by the International Monetary Fund, the Center on Global Economic Governance at Columbia University, and the OCP Policy Center, under the title, "The Energy Transition, Nationally Determined Contributions, and the Post-COP21 Agenda." The different contributions to this conference volume reflect the latest scholarship on the economics and politics of climate change mitigation, the costs and benefits of the energy transition, and how climate agreements could be strengthened, with a specific focus on the goals of the COP21 Paris Agreement.

By bringing these different contributions together, this book aims to synthesize the key insights that emerged from the latest economic research on climate change, making them user-friendly for policy makers and scholars who wish to get closer to the frontier of climate change economics. The book covers a wide range of issues, from fossil fuel supply- and demand-side policy interventions to accelerate the energy transition, carbon taxes, the enforceability of climate agreements, intergenerational fairness considerations, and the challenge of evaluating the well-being of future generations, to the role of financial markets in incentivizing climate change mitigation and financing the energy transition. The common denominator of all the chapters is an analytical framing that begins with the existence of a market failure, the absence of a price on carbon emissions. This basic market failure calls for welfare-improving policy interventions that can take multiple forms, starting with the taxation (or regulation) of carbon emissions, and extending to complementary policies such as the subsidization of renewable energy, research and development in renewables, energy efficiency and conservation, and the disclosure and measurement of carbon risk exposures so that they can be managed.

Another thread linking the chapters is the recognition that the huge global pool of savings that has accumulated as a result of the high growth rates of emerging market economies in the past two decades and the buildup of precautionary savings following the financial crisis of 2007–2009 provides both a financing opportunity for the energy transition and investments in sustainable infrastructure, and a challenge of removing the institutional bottlenecks that prevent the flow of funds away from low-yielding treasury securities to higher-yielding infrastructure assets. Beyond exploring more traditional policy interventions, this book argues that more coordinated action toward bridging the gap between global savings and investment needs is a key step in building a more sustainable global economy.

Both academic economists and practitioners from the public and private sectors have contributed to this book. Some of its chapters will surprise newcomers to the field in revealing how seriously public and private organizations—both country-level players and multilateral institutions such as the International Monetary Fund—are taking global climate risks and are planning for the future.

The chapters tackle the four main themes around which the book is organized: (i) The energy transition and its consequences; (ii) carbon pricing and the implications of uncertainty around the pace of global warming and tipping points; (iii) implementing climate agreements; and (iv) finance and sustainable infrastructure.

The Energy Transition and Its Consequences

Climate goals cannot be achieved without a significant and rapid energy transition away from fossil fuels. Indeed, at the heart of the implementation of the Paris Agreement is the move away from fossil fuels toward clean energies to power the global economy. As urgent and dramatic as this transition sounds, the good news is that there are increasingly credible opportunities associated with renewable energy technologies. While the energy transition is arguably at an early stage, with important differences across countries, it is nonetheless at a critical juncture. To avoid the irreversible consequences of climate change induced by greenhouse gas emissions, the energy transition must widely and firmly take root at a time when fossil fuel prices are likely to stay low for extended periods of time. The first part of this book assesses where we stand in terms of the energy landscape and a variety of supportive policies including, research and development directed toward meeting particular climate targets. Underpinning these analyses are integrated climate-economy assessment models, which are constantly being extended and refined by scholars.

Philippe Benoit of the International Energy Agency provides an overview of the challenges and key steps necessary to reduce energy-related greenhouse gas emissions consistent with the Paris Accord's goals. Specifically, he explores the changes that are necessary over the medium to long term in the technologies of energy production and consumption. A particularly striking observation in his chapter is that demand-side interventions and state-owned enterprises are likely to play a critical role in the energy transition.

Rick van der Ploeg of the University of Oxford articulates how the concept of "safe carbon budget" could offer a robust and pragmatic approach to climate policy. The safe carbon budget is given by the maximum amount of carbon emissions such that average temperatures do not exceed a tolerable level. Associated with the notion of safe carbon budget is an optimal carbon price, which governs how quickly fossil fuel energy use is replaced with renewables energy and by how much carbon emissions are abated. In contrast to the unconstrained optimal carbon price highlighted in many economic studies, the safe carbon budget does not directly depend on ethical considerations about how to weigh the welfare of future generations or how willing current generations are to sacrifice consumption to curb future economic damages from global warming.

Getting the Price of Carbon Right

Basic economic principles suggest that the way to curb carbon emissions is to tax them: to impose a carbon price on carbon emissions. Since carbon emissions spread evenly in the atmosphere, any local emission imposes the same global harm on the world. This implies that a common carbon price should be imposed around the world. But how high should this price be? To determine the correct price requires establishing the present discounted value of the harm created by carbon emissions. Since there is a substantial lag between the time when emissions take place and the time when temperatures rise, the harm caused by carbon emissions is not immediately felt, and the worst harm may only manifest itself decades from now. This means that to determine the present cost of emissions, a complex valuation analysis is required, which is obviously sensitive to the discount rate used to apply a present value to future harm. In other words, a key step in determining the right level of carbon price is selecting the right discount rate. The choice of discount rate, in turn, requires determining the nature of the risk exposure with respect to the future harm caused by carbon emissions. What are the worst-case scenarios, how is future harm tied to growth and consumption growth risk? The answers to these questions govern what risk premium to apply to the long-term safe interest rate to discount the future cost of current emissions. Many countries have embarked on important reforms to determine what discount rate to use and how to quantify the climate risk associated with greenhouse gas emissions.

Even when the right level of carbon prices has been determined, the inevitable next question is: How should the carbon tax proceeds be redistributed so as to equalize the cost of abatement across countries? Or, how should the tax be adjusted to the local capacity of each country to bear the abatement cost? Carbon pricing has important redistributive implications, both across and within countries. These implications may require gradual implementation of a carbon tax, complemented with mitigation and adaptation measures that shield the most vulnerable. All that being said, the current low level of fossil fuel prices provides an opportune moment to eliminate fossil fuel energy subsidies and to introduce carbon prices.

Christian Gollier of the Toulouse School of Economics argues that fighting climate change yields short-term collective costs, thereby creating a political problem for benevolent decision makers who support an ambitious international agreement. This obstacle means that the possibility of putting in place a globally efficient carbon pricing policy remains a remote and somewhat unrealistic prospect. Gollier's chapter suggests that the social cost of carbon, that is, the globally optimal carbon price, is around $40 per metric ton of CO_2. In other words, it is socially desirable to implement any step to reduce carbon emissions as long as the present value cost of that investment is below $40 per metric ton.

Ian Parry of the International Monetary Fund distills recent analyses of the efficiency of fossil fuel prices, focusing on the G20 group of member countries, especially China (by far the largest aggregate emitter). He shows how the efficient fuel price should include a number of elements, in particular the supply cost, the environmental costs—both global and domestic—and the general sales taxes applied to final goods produced with the fossil fuel inputs. Equally important is how the harm caused by emissions is calculated. Parry shows that among the substantial benefits of emission reductions are the health benefits from lower local air pollution.

Katheline Schubert of the Paris School of Economics asks why the basic principle that the same global harm should be taxed equally everywhere is not applied to implement the same carbon tax around the world. She points out that international differences in fuel taxation are huge, and may be justified by different local negative externalities that taxes must correct, as well as by different preferences for public spending. An important result is to show that the uniformity of the carbon price across countries could still hold in a second-best world. Nevertheless, if lump-sum transfers between governments are impossible to implement, international differentiation of the carbon price is the only way to address equity concerns.

Ted Loch-Temzelides of Rice University argues that in the presence of climate uncertainty, economic decisions and policy ought to pay particular attention to "worst-case scenarios." In dynamic economic modeling, this dictum can be operationalized using robust control. This approach has several implications for policy, as well as for the optimal energy transition. As a rule, mindful of worst-case scenarios, these models prescribe taking a more cautious approach. A more sizable optimal carbon tax on emissions is needed to ensure that the excessive damages of the worst-case scenarios are avoided. This prescription is in stark contrast with existing policies around the world, which in extreme cases subsidize fossil fuel instead.

Climate Agreements

The third major topic of this book is the credibility of climate agreements. Emission reduction promises are subject to the well-known "free-rider" problem. Better to let others do the costly abatements in carbon emissions. The enforceability of climate agreements thus rests on reputation, accountability, and sanctions. The big bet of the

Paris Agreement is that reputation concerns, the pressure of public opinion will go a long way in inducing countries to honor their commitments. This bet has been put to the test by the Trump administration, which has decided to pull the United States out of the Paris Agreement. The United States faces no sanctions for defaulting on its promises. One could have concluded that this would be a fatal blow to the Paris Agreement. But this is not what has happened so far. Both within the United States, with mayors of major cities and governors of some of the biggest states responding that as far as they are concerned this decision changes nothing for their plans to implement the Paris commitments, and around the world, with no other ratifying country following the United States in pulling out of the Paris Agreement, the reaction has been that this move by the Trump administration if anything strengthens their resolve to carry out what they promised.

Yet, as heartening as the world's reaction has been to the Trump decision, one cannot entirely ignore the concerns of many commentators about the Paris Agreement that the NDCs may in the end not be implemented because no explicit economic incentives to complete the NDCs were included in the agreement. An important open question remains—How to create incentives other than reputation concerns to get countries to honor their promises to reduce carbon emissions? This will become especially true when the abatement costs become significant and when more drastic measures have to be taken to curb emissions. This is also true in particular for the commitments by developed countries to financially support developing countries' transition to renewable energy.

Bård Harstad of the University of Oslo makes several specific suggestions to improve climate treaties. In particular, two policies are not subject to binding commitments in the Paris Agreement, namely countries' investment levels in "green" technologies and countries' extraction levels of natural resources. Climate treaties should be reviewed and revised over time, and the types of technologies that can make the commitments most credible need to be identified. Supply-side climate policy is especially helpful and even necessary as a complement to traditional demand-side policies.

William Nordhaus has famously put forward the idea of a "climate club" to avoid the free-rider problem. Martin L. Weitzman of Harvard University builds on Nordhaus's proposal and suggests that a "climate assembly" could be established to decide on the flexible adjustments to climate agreements that become necessary as circumstances change. Weitzman argues that it is difficult, if not impossible, to resolve the global warming free-rider externality problem by negotiating over many different national emission quotas. In contrast, negotiating over a single internationally binding minimum carbon price (the proceeds of which are domestically retained) counters self-interest by incentivizing countries to internalize the externality. Each country is then prepared to shoulder the extra cost from a higher emissions price if it is counterbalanced by that country's extra benefit from inducing all other countries to simultaneously lower their emissions in response to the higher price.

Maurice Obstfeld of the International Monetary Fund (IMF) points out that the IMF has long highlighted the macroeconomic risks associated with climate change. He argues that the IMF should further help countries to adapt the energy sector of their economies and to meet their emission reduction promises in the Paris Agreement. There are several important reasons that justify a role for the IMF in addressing climate change, including avoidance of economic coordination failures.

Ujjayant Chakravorty of Tufts University, Carolyn Fischer of Resources for the Future, and Marie-Helene Hubert of the University of Rennes argue that the success of the Paris Agreement depends critically on whether China is able to meet its stated environmental goals by the year 2030. They suggest that it will likely not be easy to meet these targets without putting some form of carbon pricing in place. If the benefits from large cost reductions in solar and wind energy continue to be realized, however, both targets could be met by 2030, and at a reasonable carbon price. A critical question is regulatory reform of the electricity sector in China and the role of emissions trading markets more generally.

Finance and Sustainable Infrastructure

The fourth major topic of the book is the role of financial markets in accelerating the energy transition and financing sustainable infrastructure. For many developing countries, financial aid may be necessary to facilitate the clean technology imports necessary to ensure their participation in the energy transition. This financial aid could among other things help offset the countries' transitional costs associated with removing carbon subsidies and levying positive carbon taxes. In this vein, the Green Climate Fund—a fund within the framework of the United Nations—was founded as a mechanism to assist developing countries in their adaptation and mitigation policies. The Fund is intended to be the centerpiece of efforts to raise climate finance to $100 billion a year by 2020. The private sector plays a central role in providing this funding for "green infrastructure."

Thierry Déau and Julien Touati, (both of Meridiam) explore the challenges of channeling savings, primarily collected by institutional investors (life insurers and pension funds), toward the energy transition. To address the scale and urgency of the sustainable infrastructure financing challenge, infrastructure finance will need to reinvent itself in the coming years, both in the developed and in the developing world. Initiatives that are badly needed include setting up incentive compatible environments to secure much-needed investments in low carbon infrastructure projects.

Jean Boissinot of the Direction Générale du Trésor, and Frédéric Samama of Amundi explore the role that financial markets can play in creating incentives toward the low-carbon transition. They argue that several developments have been instrumental to the process of mainstreaming green finance: (i) a better understanding of the implications of climate-change–related developments for financial institutions; (ii) financial innovations to address scalability; (iii) allowance for time horizon and complexity issues; and, (iv) a forward-looking attitude within the financial sector. They argue, however, that just as governments cannot take on the climate challenge alone, nor can financial markets and the private sector. To fully leverage the power of financial markets, governments need to be prepared to take a more proactive role by mandating better disclosure of carbon footprints, by sometimes offering limited guarantees, and by taking on risks that investors are not always well equipped to hold.

Conclusion

As we write this introduction, the political landscape has shifted, with the United States seeming to abandon the COP21 Paris Agreement. It is hard at this stage to assess the implication of the U.S. decision.

At the same time, it is important also to avoid overstating the likely fallout from that decision. In contrast to the U.S. administration, several states and large corporations have publicly stated they will still abide by the Paris Agreement. Europe and Asian countries including China seem intent on carrying the torch and moving forward even without the United States on board. The Paris Climate Agreement has been a success for the very reason that it adopted a more inclusive strategy. The Paris Agreement in particular recognized the critical role of developing and emerging market countries as well as the private sector. So, while the U.S. decision is a setback, it may only be temporary and the United States may well rejoin the international community in fighting climate change in the future. Private sectors and local government may play an important role in filling gaps in the meantime. All in all, the nature of the Paris Agreement may make it rather resilient. Notwithstanding the U.S. decision to withdraw, technological progress in renewable energy, enhanced energy efficiency, higher awareness of climate change risks (including in the corporate sector), and the advent of green finance are major forces that will support the Paris Agreement.

Further research and dialogue is essential however to understand the potential distributional consequences of climate policies. Indeed, while the Paris Agreement addressed quite successfully the "North–South" divide between rich and poorer countries, the outcome of the 2016 U.S. presidential election is a harbinger of a coming backlash against international trade and multilateral international agreements that are seen as engines of domestic economic dislocation. Distributional issues within countries as opposed to between countries are equally important and cannot be neglected. A constructive critique of mainstream approaches to addressing climate change is that they have failed to adequately incorporate within-country distributional concerns arising from the implementation of carbon taxes and the transition away from fossil fuels. From a policy standpoint, to avoid leaving parts of the population behind, human investments including retooling and retraining should be considered, as some have recommended in the face of other technological changes and globalization.

PART ONE

The Energy Transition and Its Consequences

CHAPTER ONE

Reducing Energy Greenhouse Gas Emissions to Meet Our Climate Goals

An Overview

PHILIPPE BENOIT

Introduction

The Paris Agreement reached at COP21 in December 2015 (which went into effect on November 4, 2016) has set a new and ambitious climate target of limiting global temperature increase to "well below" 2°C (namely 3.6°F) above pre-industrial levels, a significantly lower target than the previous 2°C threshold. The agreement represents the surprisingly successful culmination of an international diplomatic effort to establish a global framework for all countries to work to reduce greenhouse gas (GHG) emissions and their pernicious impacts. It is noteworthy not only because of its ambition but also because virtually every nation in the world (developed and developing) submitted their own program of policies and actions to limit their GHG emissions and to manage the adverse impacts of climate change; each country set out its program in its intended nationally determined contributions ("INDC").

The focus has now moved to implementing these programs to reduce emissions, notwithstanding the recent decision of the United

States to withdraw from the Paris Agreement. Change on the ground in how we produce and consume energy is the climate challenge of the next five to ten years (and beyond). While important actions have already taken place (reflected, for example, in the massive increase in the deployment of renewables over the last decade), significantly more action is needed to achieve the goals of the Paris Agreement. Their achievement will require extensive efforts to reduce energy sector carbon emissions across a variety of energy technologies by mobilizing a wide array of approaches that will require actions by governments, private businesses and households. The inauguration of a new U.S. administration in January 2017 and the related shifts in U.S. positioning and policies regarding the Paris Agreement specifically and the climate change effort generally have added another layer of complexity and difficulty to the effort to limit global temperature increase to 2°C or less above pre-industrial levels. This chapter provides an overview of the challenges and key actions on the path to success, while also shedding light on some important, yet underappreciated tools in this effort, notably the roles of demand-side drivers and state-owned enterprises.

1. Mobilizing a Portfolio of Clean Energy Technologies to Limit Temperature Increase

International Energy Agency (IEA) analysis has consistently demonstrated that there is a cost-effective pathway to reduce energy sector emissions to a level that is consistent with limiting temperature increase to 2°C while supporting continued global economic growth. In its "2DS" (the two-degrees scenario), the IEA sets out a portfolio of energy investments and deployments that yields a GHG emissions pathway for the sector consistent with limiting the global temperature increase to 2°C (IEA 2016b). This is done through a combination of increased investments in energy efficiency, renewables, nuclear power generation, carbon capture and storage and fuel switching (figure 1.1). These investments enable the energy sector to expand to support a global gross domestic product (GDP) level that is over three times as large as today's (figure 1.2) while reducing emissions to a level in 2050 emissions that is less than half of today's level (namely, less than 15 gigatonnes [Gt] in 2050).

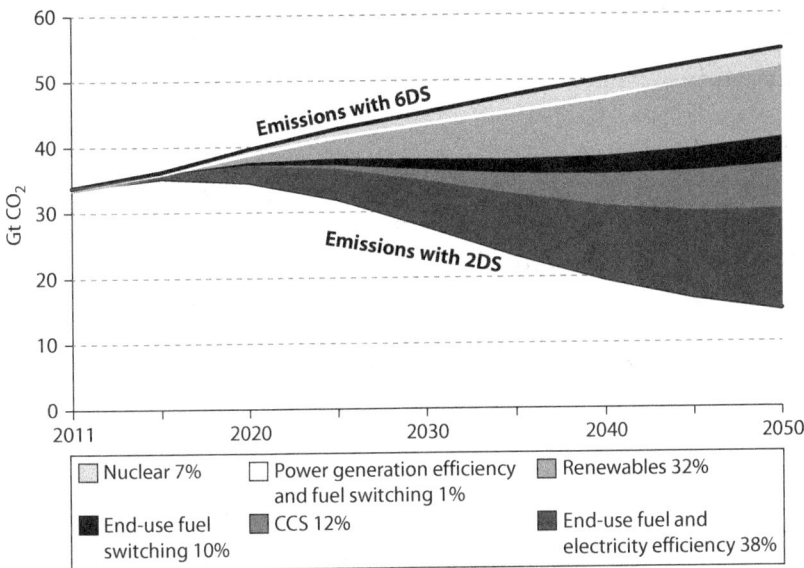

Figure 1.1 Portfolio of energy technologies under ETP 2DS modeling. *Source*: IEA (2016b).

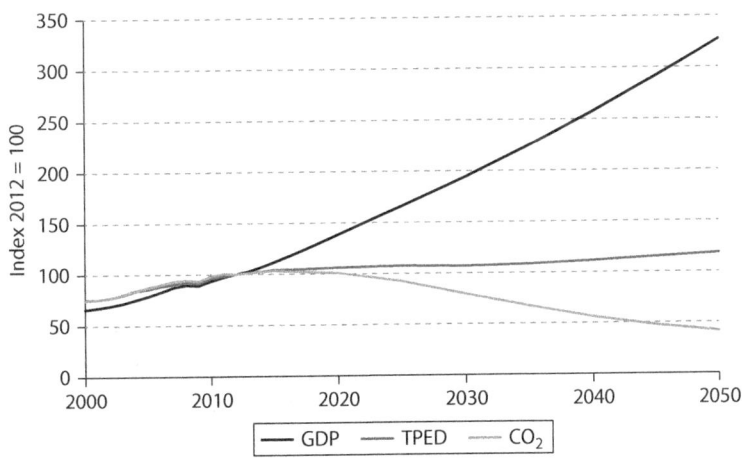

Figure 1.2 Evolution of GDP, total energy demand, and CO_2 emissions under the 2DS. *Source*: IEA (2014a).

Although the 2DS sets out a pathway to dramatically reduce emissions, the new country policies set out in the INDCs will generate substantially higher emissions consistent with a global temperature increase of 3°C or more (which is a substantial improvement over previous trends that were more consistent with a 4°C or even a 6°C increase [IEA, 2014a]). At the same time, the Paris Agreement has established a climate target that is now "well below" 2°C. Implementing the new policies set out in the INDCs, while exploring opportunities for greater reductions, is a challenge facing the international community.

1.1 The Two Keys: Delinking GDP from Energy Supply and Decarbonizing That Energy Supply

Global GDP growth over the last several decades has been driven—and accompanied—by increased energy consumption, which in turn has generated increased GHG emissions. Breaking the link between GDP growth and energy sector GHG emissions involves two basic steps: delinking GDP growth from energy use and reducing the carbon intensity of our energy supply.

Investments in energy efficiency and structural changes in the economy enable GDP under the 2DS to increase to a three-fold level in 2050 while total primary energy demand ("TPED") increases by less than 20 percent—see figure 1.2. As a result, the energy intensity of the economy (energy/GDP) drops by over 60 percent by 2050. The roles of energy efficiency and structural change in driving this reduction are described in greater detail in section 3.

The deployment of renewables substantially reduces emissions by replacing fossil fuel use, thereby decreasing the carbon intensity of the energy supply (CO_2/energy). Under the 2DS, carbon emissions are more than halved by 2050 (figure 1.2) and the carbon intensity of the energy mix drops by over 60 percent. This carbon intensity of the energy mix is captured in the "energy sector carbon intensity index" (ESCI), an indicator developed by the IEA to track the evolution in the amount of carbon emitted per unit of energy supply. In comparison to the need for the ESCI to drop by over 60 percent by 2050 under the 2DS, the ESCI in actuality has been surprisingly stable and rigid, notwithstanding the dramatic increase of the last decade in the deployment of renewables (in part because of a countervailing increased in coal consumption). In fact, this "fixity" of the ESCI extends back to 1990 and earlier, which arguably illustrates the difficulty of the challenge to dramatically transform our energy mix.[1]

1.2 The Nationally Determined Contributions: Stuck at Well Above 3°C?

As part of the preparation for COP21 held in Paris in December 2015, nearly every country in the world (over 180) submitted plans to reduce their GHG emissions under their INDCs, which have now been converted into national determined contributions—"NDCs"—under the terms of the Paris Agreement. These NDCs represent the first effort by virtually all the countries of the world (and notably both developed and developing) to set out plans to control their GHG emissions as part of a global effort.

The IEA analyzed the potential impact of these plans on global energy sector emissions under its "INDC Scenario" (IEA 2015a; see also figure 1.8 later in this chapter). There were two important findings from the INDC Scenario: (i) that the NDCs do achieve a plateauing of energy sector CO_2 emissions by 2030 (namely, a minimal rate of increase in emissions at that time, which contrasts with the historically large increases of the last several decades), but (ii) that the projected emissions are consistent with a pathway that will generate a temperature increase of about 3.5°C in the longer term (IEA 2015a)—which is well above 2°C and a far cry from the "well below 2°C" climate target of the Paris Agreement. As a result, significant additional efforts will be required beyond what is set out in the NDCs to meet the climate goal of the Paris Agreement. The drafters of the Paris Agreement were aware of this shortcoming and provided in the agreement for a process that would promote gradually increasing policy ambition and action, with the current NDCs representing a first, albeit important, step on a pathway to well below 2°C.

1.3 Trying to Move Beyond 2°C to 'Well Below' 2°C Requires a Change in Perspective

As described above, one of the most notable aspects of the Paris Agreement is that it sets a new target of well below 2°C, establishing significantly greater ambition than the 2°C target which dominated much of the climate discussion and analysis to this point.[2] The "well below" 2°C target embodied in the Paris Agreement generates two new and distinct challenges. First, what does the new well below 2°C threshold mean? Does it mean 1.6°C, 1.7°C, 1.8°C, or something else? The second is, irrespective of the new threshold, how can it be achieved? What is the portfolio of actions required to achieve this lower level of emissions as compared to the well-established 2DS emissions profile? While negotiators (surely with the input of a variety of stakeholders, including their governments, scientists, and others) work to define what constitutes well below 2°C, the IEA and other energy, economic, and climate agencies and organizations will look to establish ways to achieve an emissions pathway that is indeed consistent with limiting temperature increases to well below 2°C.

Developing an emissions pathway consistent with the new well below target will require somewhat of a shift in perspective from the 2DS and other two-degree scenarios (such as the IEA's 450 ppm scenario presented in the World Energy Outlook (IEA, 2015b) that have dominated the discussions to date. The 2DS reduces emissions, but it does not eliminate them. As the well below target requires fewer GHG emissions than set out in the 2DS pathway, it is no longer simply a question of reducing emissions as provided in the 2DS, it is also necessary to eliminate some of those emissions that remain in the 2DS, namely to lower the 2DS pathway. As a result, in addition to the wedges that reduce emissions down to the 2DS pathway (see, for example, the wedges between the 6DS and 2DS lines set out in figure 1.1), some of the emissions remaining under the 2DS (namely those that sit below the 2DS line in figure 1.1) will need to be eliminated. Figure 1.3 shows the composition by energy subsector (power, transport, industry, etc.) of the emissions sitting below the 2DS line that will now also need to be tackled to move to well below 2°C.

The subsectoral composition of actual GHG emissions in the 2DS has industry, power, and transport accounting for over 80 percent of these emissions over the period from 2015 to 2050 (figure 1.4). However, as illustrated in figure 1.3, power sector emissions represent only a small portion in 2050 as compared to industry and transport. Accordingly, efforts to reduce emissions to a level well below 2°C will require greater efforts on industry and transport sector emissions.[3] However, as power still represents nearly 30 percent of the emissions in the 2DS over the period, and today accounts for about 40 percent of annual energy sector emissions, efforts to accelerate decarbonization

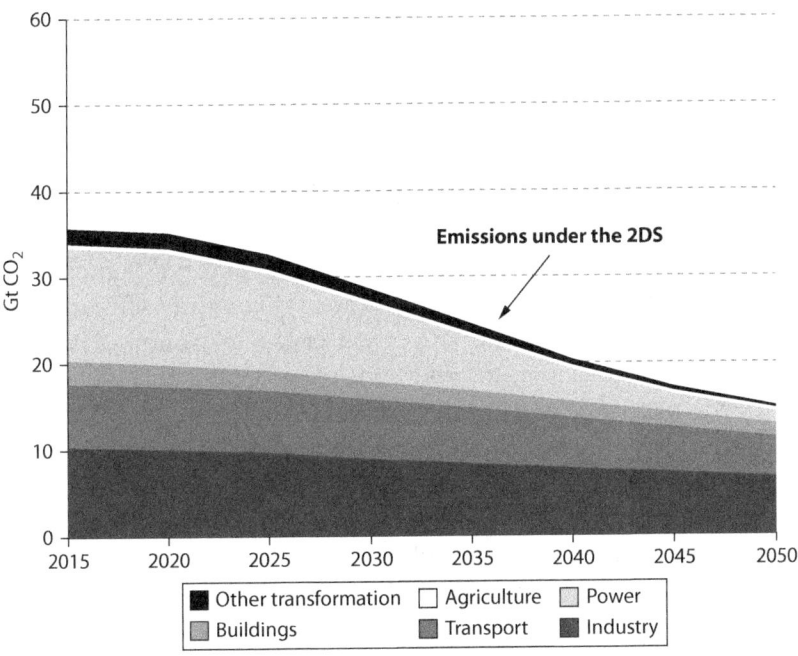

Figure 1.3 Achieving the "well below 2" target: Emissions by sector "below the 2DS line." *Source*: IEA (2016a).

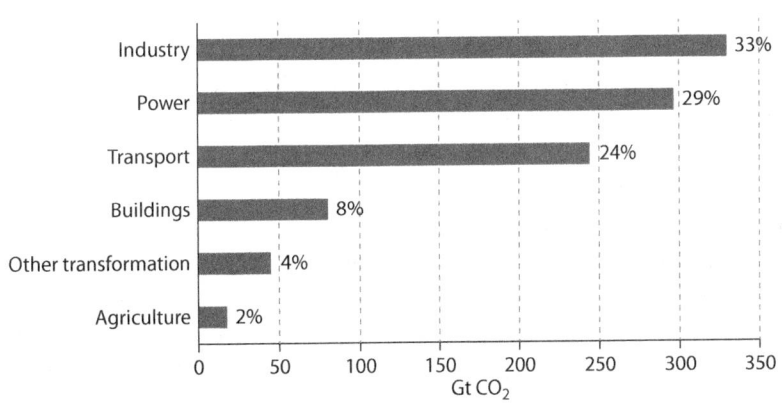

Figure 1.4 Sectoral breakdown of cumulative emissions through 2050 in the 2DS. *Source*: IEA (2016a).

of the power sector will also support efforts to move well below 2°C and still remain key to achieving even 2°C.

2. Electricity: At the Center of a Low-Carbon Energy Transition

As noted earlier, electricity today generates 40 percent of energy sector emissions, about twice as much as any other energy subsector, and consequently is at the center of current efforts to reduce GHG emissions. Renewables and nuclear power generation, together with energy efficiency measures, provide viable cost-effective tools to reduce electricity sector emissions. There are three key dynamics in this regard: (a) reducing the use of high-carbon fuel sources for generation; (b) expanding the use of low-carbon resources to substitute for the aforementioned reduction in high carbon and to support the expansion of "zero-carbon" energy sources (which will be needed given the projected increase in overall energy demand even in a low-carbon pathway—as discussed in section 3); and (c) optimizing the demand for electricity through energy efficiency and other measures.

2.1 Coal Is the Elephant in the Room . . . But It Can Hide Gas Which Also Presents Challenges

Coal power generation produces over 9 Gt of CO_2 emissions, larger by itself than the combined energy emissions of the United States and the European Union (EU); but coal-fired power plants also provide about 10,000 TWh, powering businesses, households, and social services worldwide. The challenge presented by coal in power (and in other uses) is that it is currently a high-carbon source, in particular relative to the decarbonization needed to reduce energy sector emissions. For example, even what are referred to as "efficient" coal power plants are estimated to emit above 700 gCO_2/kWh, while overall electricity emissions need to fall below 300 gCO_2/kWh by 2030, and to about 100 g/kWh by 2040 (see IEA 2016a,b). As a consequence, absent the use of carbon capture and storage and/or major technological improvements to reduce CO_2 emissions, it is difficult to accommodate significant coal power generation with the level of reduced emissions required to limit temperature increases to 2°C or well below 2°C. IEA modeling addresses coal's GHG emissions through three basic actions: (i) coal generation is largely replaced by renewables and gas, (ii) over 3000 TWh of coal generation is maintained through the use of carbon capture and storage technology, and (iii) the need for additional power is limited through energy efficiency measures (although the total demand for electricity actually grows in absolute terms by 2050 as it powers continued GDP growth).

One major challenge is that nearly 2000 GW of coal plants are currently in operation or expected to be commissioned over the next several years. Dealing with these "incumbent" plants will be one of the major practical, let alone political and economic, challenges facing efforts to decarbonize the electricity sector. Carbon pricing is one mechanism to incentivize the shift from coal, but experience shows that regulatory and other actions will be required, especially as expectations for carbon pricing remain modest.[4]

Gas, by comparison, is a significantly less carbon-intensive power source. For example, combined cycle gas turbines have been estimated to emit about $350gCO_2/kWh$. As a result, replacing coal generation with gas generation reduces emissions, as has occurred in the United States. However, given the increasing constraints on emissions of the 2°C target, even gas becomes a relatively high-carbon emitter over the medium term. Specifically, combined cycle gas turbines (CCGTs) are estimated to emit more per kWh by 2030 than what is required under the 2DS; the use of carbon capture and storage (CCS) can serve to reduce these emissions to well below 60 g/kWh, thereby extending gas's role as a low-carbon power source. Consequently, under the 2DS, while gas generation without CCS increases from today's levels of about 5000 TWh to nearly 6000 TWh in 2030, it plateaus thereafter before dropping dramatically to 3500 TWh in 2050.[5]

2.2 Renewables and Energy Efficiency to the Rescue

Under the 2DS, energy efficiency and renewables provide the bulk (about 70 percent) of the needed emissions reductions across all energy uses (see figure 1.1), and dominate decarbonization efforts in electricity. The contributions of these two clean technologies are amply described in a variety of publications by the IEA and others, including the *Energy Technologies Perspectives 2016* (IEA 2016b). The growth in renewables over the past decade has been remarkable, driven in particular by innovation and reduced costs, but challenges await—in particular to meet the ambitions of the Paris Agreement. These challenges include integrating high shares of renewables in our electricity systems and expanding the use of renewables in transport and industry/heat. Surprisingly for many, energy efficiency is consistently modeled as the primary contributor to lower energy sector emissions (nearly 40 percent as compared to about 30 percent for renewables [IEA 2014a]), with a large untapped potential for additional profitable investments (see, e.g., IEA 2016a). However, much public and policy attention continues to remain fixed on supply-side solutions such as renewables and fuel switching from gasoline to electricity for cars and from coal to gas for power generation, while overlooking the importance of demand-side drivers. In part to help to address this oversight, this chapter includes a section dedicated to demand-side drivers (discussed in section 3).

2.3 Nuclear: The Problem Child Among Clean-Energy Technologies

Under the 2DS, nuclear power generation provides far less than 10 percent of emissions reductions through 2050 (see figure 1.1). However, because it can provide a low carbon source of energy, it is a major element in enabling electricity to become the largest source of total final energy consumption by 2050 under the 2DS (exceeding oil) as its contribution to power generation increases from 11 percent today to 17 percent in 2050. This increase in the share of nuclear generation occurs even while its share of capacity remains level, reflecting its ability to provide stable base load power generation. But significant safety and environmental concerns remain, as well as a fundamental unease among much of the public with this technology. Moreover, its acceptability at a global level is volatile and unpredictable, and is affected both by events at a single plant (as illustrated by Fukushima) and broader ongoing concerns surrounding the disposal of spent nuclear fuel. The ability to deploy nuclear power generation may ultimately be driven more by non-climate concerns rather than its role as a low-carbon source of power generation.

3. Importance of Energy Demand in Emissions Reduction Efforts[6]

As noted earlier, energy emissions are the result of two factors: the carbon intensity of the energy supply and the amount of the supply.

The natural extension of this relationship is that actions that reduce the demand for energy supply in turn can reduce the amount of energy sector emissions. This section addresses the role that energy efficiency, structural changes in the economy and targeted energy conservation measures can have on energy demand and, by extension, on energy sector emissions. Achieving the heightened ambition of the Paris Agreement will require greater attention from governments on these various demand-side drivers.

3.1 Importance of Energy Demand in Emissions Reduction Efforts

As noted above, it is energy efficiency, and not renewables, that contributes in the IEA's analysis the largest share of emissions reductions toward limiting temperature increase to 2°C (see, for example, figure 1.1), revealing the importance of demand-side interventions. One of the major reasons for this is that the level of demand for energy is, unsurprisingly, correlated with the level of emissions. This is well illustrated by the *World Energy Outlook 2015* (IEA 2015b) analysis of different demand and emissions scenarios (see figure 1.5).

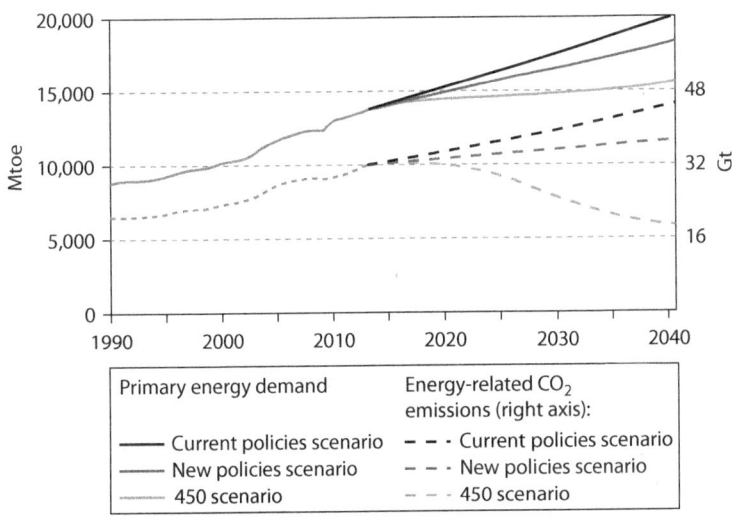

Figure 1.5 Energy demand and related CO$_2$ emissions by scenario. *Source*: IEA (2015b).

3.2 Why Energy Efficiency, Structural Change, and Energy Conservation Measures Are Relevant

There are two principal drivers that can alter energy demand without adversely affecting GDP growth:

- At the micro level, the efficiency with which equipment uses energy to produce output, namely energy efficiency
- At the macro level, the mix of economic activity across sectors and within them, given the different energy requirements of different activities, namely the impact of the economy's structure

There are also numerous programs designed to encourage consumers to reduce energy consumption, i.e., conserve energy, while maintaining their standard of living. These programs encourage consumers to adopt, for example, more informal business attire during the summer to reduce the demand for air conditioning, or to switch to more energy-efficient modes of transport. These conservation actions are described in more detail in the following section.

Efficiency effects, structural effects, and energy conservation programs can operate to avoid energy use even in the context of rising demand for energy services—i.e., producing a notional energy savings relative to a counterfactual, and similarly reduced GHG emissions relative to that counterfactual. This is particularly relevant for many emerging economies in which energy consumption has risen substantially since 2000 and is expected to increase further because of population and economic growth (IEA 2015b). Accordingly, these demand-side levers can operate to reduce GHG emissions relative to a counterfactual even in a country context in which actual energy consumption rises in absolute terms.

3.3 Energy Efficiency: The Largest Contributor to Emissions Reductions

Energy efficiency measures are among the most cost-effective actions that can be deployed to reduce emissions in the short, medium, and long term. In the 2DS set out in the *Energy Technologies Perspectives 2016* (IEA 2016b), energy efficiency improvements in end uses make the largest contribution (nearly 40 percent) to global emissions reductions through 2050, compared with the business-as-usual "6DS" scenario (IEA 2016b); renewables (the second largest contributor) provide about 30 percent of reductions. Energy efficiency improvements typically systematically reduce demand.[7] Energy efficiency's contribution to emissions reductions results from this avoided energy consumption: just as there are emissions typically associated with energy consumption, there are corresponding notional emissions that are avoided with reduced consumption, producing a reduction in energy sector emissions relative to the reference case.

The energy efficiency contribution to the 2DS is the result of substantial efficiency gains in all end-use sectors through the implementation of measures such as higher fuel economy standards in the transport sector; the adoption of highly efficient technologies to provide process heat and steam in the industrial sector; and improved efficiency standards in appliances and other residential products and industrial motors, all of which act in relative terms to reduce the demand for electricity and other energy sources. The impact of these measures on energy-related emissions is large as they reduce energy consumption under the 2DS relative to the 6DS business-as-usual reference case: total primary energy demand is about 277 EJ lower in the 2DS than in the 6DS, equivalent to nearly half of today's total global energy demand.

To date, the avoidance of fuel consumption[8] from energy efficiency improvements has significantly reduced GHG emissions. For example, the avoided emissions in 2015 in IEA member countries from energy efficiency improvements made since 2000 was 1.6 $GtCO_2$ (IEA 2016c). Cumulative avoided emissions since 2000 from these improvements are 13.2 $GtCO_2$—more than the combined emissions of all IEA countries in 2015 (see figure 1.6).

3.4. Structural Changes Can Affect Energy Demand

Structural changes in the economy and within its sectors can also reduce energy demand. For example, the IEA estimates that structural shifts toward less energy-intensive service sectors in IEA member countries since 2000 have lowered demand (IEA 2016c), offsetting in part the increase in TFC that should have resulted from increased economic activity. Structural modifications since 2000 have reduced total

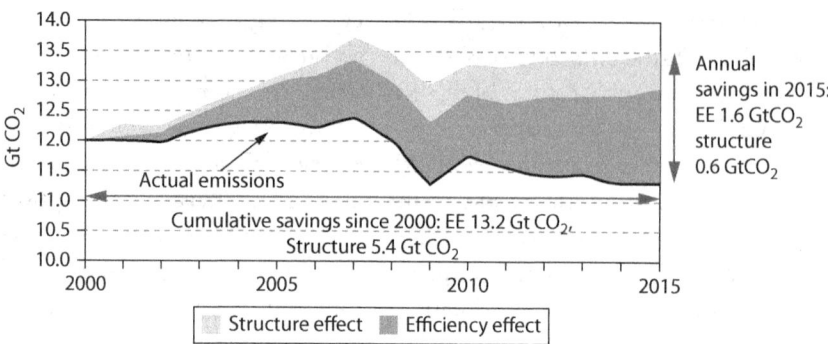

Figure 1.6 IEA member country end-use emissions savings from energy efficiency improvements and structural changes since 2000. *Source*: IEA (2016a).

final consumption (TFC) in IEA member countries by about 26 EJ in total over the 2000–2015 period (IEA 2016c). As described by the IEA, this impact is due primarily to shifts toward less energy-intensive industry and services subsectors across IEA countries (IEA 2016c). At the same time, certain structural changes have exerted upward pressure on consumption, such as the structural effects of larger homes in the residential sector and a shift to more energy-intensive modes of transport (including increased use of airplanes). However, taken in its entirety, structural change in IEA member countries since 2000 has reduced energy consumption (once again, relative to a counterfactual).

This reduction in energy consumption tied to structural changes can be converted into notional savings in emissions (figure 1.6): the structural change impact since 2000 has generated a notional cumulative savings in emissions of 5.6 $GtCO_2$ through 2015 (IEA 2016a). This structural effect in IEA member countries has been small compared to the impact of energy efficiency improvements; for example, the structural effect in 2015 from changes since 2000 has been only 3 percent, compared with a 15 percent efficiency effect from energy efficiency improvements over the same period. Correspondingly, the savings in emissions from structural changes is less than from energy efficiency.

While energy efficiency improvements typically systematically reduce demand, different types of structural changes to an economy will exert either downward or upward pressure on energy consumption. For example, energy intensity and consumption can increase in rapidly developing poorer countries as they shift from less energy-intensive agricultural activities to manufacturing. In addition, although energy efficiency improvements in one country typically produce an emissions reduction benefit at the global level, it is important to determine whether a shift in structure that leads to reduced emissions in one country is accompanied by a related increase in another country (e.g., services taking the place of manufacturing in China, with the manufacturing activity relocating to another developing country). Finally, while policy makers can influence structural change (e.g., adopting fiscal policies to promote expansion of the services sector, or policies to promote modal shifts to less energy-intensive forms of transport), much of the change in structure is driven by exogenous market forces.

China's recently adopted 13th Five-Year Plan ("13th FYP") illustrates the significant potential to affect emissions through structure. As analyzed by the IEA, the projected impact on energy demand of actions under the 13th FYP affecting structure is twice as large as the anticipated impact from energy efficiency improvements (IEA 2016c). China's strategy in the 13th FYP of extensively using structural change measures to manage energy demand sets an interesting precedent for other countries, in particular for other emerging economies that are at or nearing a similar developmental stage. It may also prove to be an important additional approach to address GHG emissions while supporting continued economic growth and poverty alleviation.

Targeting more ambitious goals beyond 2°C will require even greater attention to decoupling economic growth from energy consumption. Accordingly, structural changes will likely gain in prominence in mitigation analysis as climate ambition increases.

3.5 Energy Conservation Programs Can Also Affect Energy Demand

Many countries have programs that encourage consumers to reduce consumption by changing their behavior. An important element of the programs described in this section is that they are not adopted as emergency measures to respond to an energy supply crisis, but rather are promoted as ways to ultimately improve the participant's standard of living.[9] As a result, they are designed to create a new "normal" with respect to energy consumption behavior.

For example, the Italian government has a program that encourages employees to take the stairs rather than the elevator on certain days, which not only reduces energy consumption but also improves worker health.[10] India's Bureau of Energy Efficiency has an advertising program in which children teach parents and teachers how to save energy (e.g., opening the curtains and turning off the lights).[11] China's 13th FYP also includes various conservation elements, such as promoting behavioral changes in terms of consumer purchases and lifestyle habits (IEA 2016c). Japan's Super CoolBiz campaign encourages changes in clothing habits, to enable workers to consume less air conditioning during the summer months (IEA 2011).

"Avoid" and "shift" programs for transport can also encourage consumers to change their travel preferences to ones that require less energy, resulting in a corresponding reduction in emissions (IEA 2016b). For example, transport energy demand is "avoided" when consumers opt to walk to their workplace or can telework from their homes. "Shifting" programs target moving to more energy-efficient transport modes, for instance by encouraging commuters to take mass transit rather than their personal cars. These programs can require changes in physical infrastructure, such as creating bicycle lanes or building housing sufficiently close to office space to enable commuters to walk or cycle. The positive impacts of avoid and shift policies on emissions can be substantial, with more than one-third of the reductions in the 2DS coming from these approaches (IEA 2016b).

4. State-Owned Enterprises: Big Players in the Low-Carbon Energy Transition[12]

One of the important—and overlooked—means by which governments can promote decarbonization action is through their capacity as public shareholders of state-owned energy enterprises (including fossil fuel and low-carbon power generation companies) and of state-owned energy-intensive industries (such as steel and cement producers, and urban transit systems). Through this ownership, governments have the ability to direct or otherwise influence state-owned enterprise (SOE) investment and energy consumption patterns. In China, India, Latin America, Europe, and elsewhere many electric utilities and oil and gas producers, as well as large energy users, are state owned.

The decarbonization actions of these SOEs have often been driven by formal and informal directives, as well as financial and other incentives from their government shareholders; these constitute avenues for advancing the low-carbon transition that merit greater attention and analysis. Given the weight of SOEs in the energy sector (as emitters, operators of low-carbon generation, energy consumers, and financiers of investment), in particular in emerging economies where much of the anticipated growth in energy demand will take place,[13] it is important to explore what incentives are best suited to prompt these actors to advance low-carbon objectives.

4.1 SOEs Play a Major Role in Determining Energy Sector Emissions in a Variety of Different Capacities

SOEs account for a significant share of the global energy sector and will affect the level of emissions through a variety of differing roles. These roles include (a) high-carbon operators; (b) low- and even 'zero-carbon' energy providers; (c) major consumers of energy, including in industry and transport; and (d) funders of energy sector investments.

4.1.1 SOEs ARE DOMINANT IN HIGH-CARBON ACTIVITIES

The IEA estimates that SOEs own about 70 percent of oil and gas reserves (IEA 2014a). In the electric power sector, which accounts for over 40 percent of energy sector GHG emissions globally, SOEs own about 42 percent of fossil fuel power generation capacity.[14] SOEs owned an even larger share of the new fossil fuel generation capacity commissioned in 2015 (54 percent), of which nearly three-quarters was coal (IEA 2016b).

In many emerging economies, SOEs are responsible for a high share of energy sector emissions. In China, for example, half of energy sector CO_2 emissions are emitted by an electric power sector dominated by state-owned electricity producers and other energy companies. In India, SOEs generate over 40 percent of total thermal electricity (which produces half of India's energy sector emissions) and they also dominate in coal and oil production (OECD 2015a).

Even in Organisation for Economic Cooperation and Development (OECD) member countries where the size of the state-owned sector has declined following decades of privatization, SOEs remain influential actors in sectors of strategic importance such as energy.

Overall, the SOE portfolios of OECD member and affiliated partner countries are concentrated in energy-intensive sectors such as oil and gas, electric power, transportation and extractive industries as well as in finance (OECD 2014). For example, France's electricity sector is dominated by Électricité de France, which is 85 percent owned by the French government; Mexico's state-owned Comisión Federal de Electricidad is the principal electric utility in the country, serving over 100 million people.

4.1.2 SOEs ARE ALSO MAJOR PLAYERS IN LOW-CARBON ACTIVITIES Decarbonization requires not only reduced investments in fossil fuel use generation but also additional investments in clean energy technologies—once again, this is an area in which SOEs are active, and in certain cases are dominant. Globally, 60 percent of generation capacity in renewables and nuclear is state owned (IEA 2016a). Of the new renewable and nuclear capacity commissioned in 2015, 45 percent was state owned (IEA 2016e), with hydropower, wind, and nuclear accounting for over 90 percent of this capacity. In Brazil, China, Mexico, and elsewhere SOEs own the majority of large-scale hydropower generation, including the world's largest sites such as the Three Gorges Dam in China and the Itaipu Dam on the Brazil–Paraguay border. SOEs have also played important roles in the development of wind and solar power: Chinese SOEs, for example, have been major developers, spurred partly (in the case of wind) by government mandates requiring that a certain percentage of SOEs' new generating capacity come from this low-carbon technology.

4.1.3 SOEs ARE PRESENT AS ENERGY CONSUMERS, INCLUDING ENERGY-INTENSIVE INDUSTRIES State ownership is also important in other energy-intensive industries such as steel and cement as well as other large energy consumers, such as municipal transit systems. From the Steel Authority of India Limited and the Emirates Steel Industries to PT Semen Indonesia Tbk and China's Anhui Conch Cement Company, SOEs are important parties across industries that consume large quantities of energy or generate CO_2 emissions as part of their industrial processes (for example, in the case of cement).

When SOE industry emissions are added to those of the energy supply sector, the total GHG emissions attributed to SOEs grows. Under the IEA's analysis of SOEs, a selected group of 50 SOEs operating

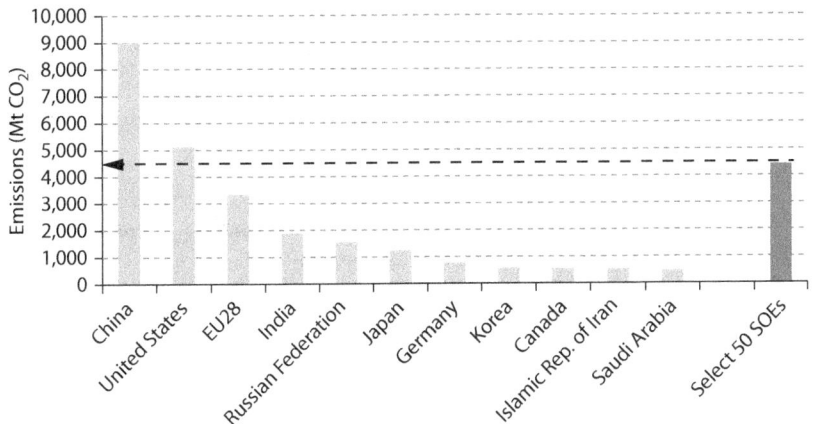

Figure 1.7 CO_2-equivalent emissions of fifty selected SOEs across various sectors and countries, compared with the top emitting countries/regions. *Source*: IEA (2016a).

Note: Country emissions are CO_2 emissions from fuel combustion for 2013 (IEA 2015d).

in power, oil and gas, iron and steel, and cement from around the world have annual GHG emissions that total over 4.5 Gt, which is higher than the energy-related emissions of every country other than China and the United States, and higher than that of the European Union and Japan combined (figure 1.7). These emissions increase further when other state-controlled entities are considered, such as the state-controlled transit systems in major cities such as New York and Paris that are major sources of energy demand and whose investment and service decisions (e.g., choice of buses) will affect energy sector emissions. The military is also a major consumer of energy in many countries.

4.1.4 SOEs ARE MAJOR FUNDERS OF ENERGY SECTOR INVESTMENTS The government is also present in the energy sector through publicly owned banks that provide financing to energy producers and users. Much of the financing for energy investments in emerging economies comes from public resources, and domestic state-owned financial institutions are likely to play an important role in financing low-carbon investments in the future (Benoit 2012).

These banks are major providers of finance for SOEs as well as the private sector. Brazil's *Banco nacional do desenvolvimento* (BNDES), for example, provided over $6.5 billion in 2014 in financing for private- and public-sector borrowers for renewables and energy efficiency (OECD 2015b).

4.2 Special Channels for Government Influence Over SOEs: Wielding Public Shareholder Power

Governments create or acquire companies to serve a variety of governmental objectives, typically economic and social development goals. As a result, SOEs are frequently motivated by factors beyond profit maximization, such as promoting economic activity, energy security, social development, electricity access, employment, and other strategic objectives. The context in which SOEs operate is generally characterized by greater political intervention than that faced by private sector counterparts, softer budget constraints, and various financial support mechanisms including access to low-cost capital and subsidized input prices (Earnhart, Khanna, and Lyon 2014); they are also often susceptible to greater political influence as a result of their ownership structure.

The ability and willingness of governments to influence and direct the corporate actions of individual SOEs often depends on the market structure and prevailing business culture in which an SOE operates—notably, the extent to which the SOE is expected to operate in a profit-driven manner subject to market forces. SOEs in different sectors often have different social and economic mandates beyond profitability, particularly in the power sector, as electricity supply has both national economic and social dimensions.

SOEs operate in a variety of sectors in many different country and market contexts, which can be distinguished by various criteria. For example, electricity, urban transport, airlines, cement, and oil are very different businesses involving distinct products and commercial settings. While some SOEs produce internationally traded goods in vibrant markets, others produce for a dedicated domestic market. Some companies operate in regulated markets while others conduct business in more competitive ones. These factors can affect how a government decides to influence a particular SOE. In certain cases, a government may exercise close control or influence over an SOE or the market in which it operates, while in others the government may remain distant giving the SOE significant commercial and operation flexibility. Although there is a wide variety of SOEs, there are certain common elements that affect SOEs to varying degrees that are pertinent to efforts to reduce energy-related emissions.

A government that is the sole or primary shareholder of an SOE can control or influence decarbonization of its SOEs through a variety of direct and indirect channels. For example, it can adopt and implement clear, consistent, and predictable policy directives to influence short-term operations (e.g., shifting electricity dispatch patterns to favor low-carbon sources) and long-term planning. These policies can be supported with informal dialogue to reinforce policy messages. A government can exercise authority to appoint (and change) senior management, which can strongly influence SOE actions (balanced with the need to avoid excessive political interference). It can influence investment patterns in specific energy technologies as a supplier/facilitator of funding for SOEs (including funding through state-owned financial institutions). Perhaps equally important, a government can provide both formal and informal signals to SOEs, which are more likely than private enterprises to follow government signaling because of their shareholding structure.

It should also be noted that governments adopt regulations and pricing mechanisms that target the economy more broadly, including private sector actors. These are not typically viewed as SOE-specific, but at times the mode of adoption and implementation of this type of action may be influenced by government ownership of key companies (e.g., simplified consultations with industry). Governments may also encourage more active SOE participation in low-carbon mechanisms (e.g., encouraging trading activity under emissions trading systems).

The exercise of public shareholder power may in some circumstances influence SOEs more than the use of price signals alone, especially when the impact of financial drivers on these enterprises is diluted by nonfinancial mandates (such as expanding energy access and other national and regional development goals). Governments can also exercise shareholder power to encourage their SOEs to engage in more technology innovation and development programs (for carbon capture and storage, for example[15]) and in international collaborative efforts.

4.3 A Need for More Analysis of SOE Dynamics

Given the central role of SOEs in generating energy sector GHG emissions and their vital role in decarbonization efforts (in reducing emissions, in providing clean energy alternatives and in supplying funding), further analysis is needed about how complementary measures beyond pricing can influence SOE action. These complementary measures are especially relevant because many SOEs—particularly in various emerging economies that are central to decarbonization—operate in contexts in which government shareholder direction may outweigh liberalized market signals. While there are common elements that characterize SOEs, their heterogeneity across sectors and countries requires a variegated approach.

5. A Role for Carbon-Pricing Mechanisms . . .
That Is Often Exaggerated

Carbon taxes and emissions trading systems are important market-based instruments that can spur the efficient reallocation of resources by properly accounting for the negative impact of carbon on our environment. Much has been written about these instruments and a great deal of policy effort has been expended to implement carbon-pricing and emissions trading system (ETS) frameworks. From the EU to China, from Sweden to British Columbia to Brazil, carbon-pricing mechanisms have been touted as a key component of emissions reduction efforts. The attention accorded these mechanisms has been even greater within academia and think tanks, and among economists (specialists that have played a visible leading role in the climate change challenge discourse). Unfortunately, notwithstanding the attention, carbon-pricing mechanisms have not delivered, in part limited by political realities that have made many countries unable and unwilling to implement frameworks that will generate the robust level of carbon-pricing incentives needed to shift resources to low-carbon alternatives. As summarized by the IEA: "After more than a decade of using carbon markets globally, . . . carbon pricing policies are not delivering their theoretical potential. Realistically achievable carbon prices in the short to medium term do not appear high enough to drive the investment and operational changes needed to decarbonise" (2016a:43).

Other mechanisms have, by comparison, produced major benefits. The feed-in tariffs for renewables (notably in Germany) helped to spur a massive expansion in the solar photovoltaic (PV) market that led to significant cost reductions (in part as China aggressively entered the PV manufacturing field). Technological improvements in the gas production field led to a domestic gas boom in the United States and a resulting drop in gas prices relative to coal prices; this in turn, resulted in the substitution of lower-emissions gas power generation for higher-emissions coal generation that has been at the heart of the recent U.S. emissions reductions. Energy efficiency regulations (including minimum energy performance standards for equipment and labeling schemes for appliances) have restrained energy consumption, while vehicle fuel economy standards have been adopted in all major economies and are helping to avoid emissions. Moreover, as discussed above, state-owned enterprises operating notably in non-liberalized markets (such as China) are major actors in the energy sector and are often driven by non-pricing incentives. Efforts to fight local pollution may in the medium term have more of a beneficial impact on reducing emissions than the modest carbon-pricing levels being produced by carbon-pricing mechanisms. In a complex world, a suite of levers is needed.

Carbon-pricing mechanisms do continue to have an important role to play, even at modest levels (IEA 2016a). For example, even moderate carbon prices in the electricity sector help to lower emissions by helping to change dispatch patterns. They can support other policy tools (such as renewable portfolio standards) by adding some economic incentives to a policy mandate. They can also play an important signaling role by making explicit to the public the economic cost of carbon emissions. The use of carbon taxes and ETSs are expanding worldwide, and may ultimately achieve their potential in efficiently reallocating resources to low-carbon investments. However, given to date the relatively limited impact on emissions reductions of modest carbon-pricing mechanisms compared to the reductions being generated by regulations and technological support mechanisms, the time has come for academic institutions and think tanks to focus more attention on these other mechanisms that are producing more tangible results.

6. The Paris Agreement and Our Climate Future Do Not Look the Same with the New U.S. Administration

The change in policies and pronouncements from the United States relative to the climate change effort can be expected to hamper efforts to limit the global temperature increase to 2°C above pre-industrial levels (let alone well below 2°C). To fully appreciate the impact of this change, it is important first and foremost to place this shift in United States policy within the climate change context and its underlying emissions challenge.

While the Paris Agreement represented an important achievement and milestone in international efforts to address climate change, the projected cumulative impact on energy sector emissions of the NDCs submitted by countries under the agreement generates only a "plateauing" of emissions by 2030 (figure 1.8) that is consistent with a 3.5°C increase; they are not sufficient to achieve the 2°C threshold, let alone the well below 2°C goal set out in the agreement itself (see discussion in section 1.2). Similarly, a leveling of emissions at current levels as occurred over the three years beginning in 2014 is insufficient to achieve the climate goal as current emissions are far above those

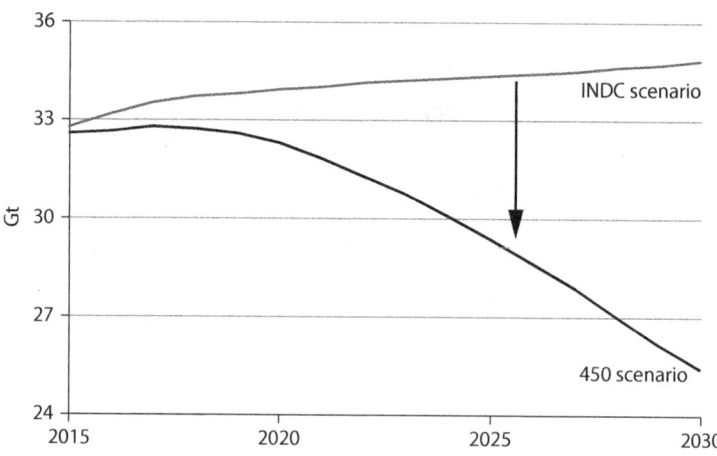

Figure 1.8 Paris Agreement "outcome" (under the INDC scenario) compared to the 2°C pathway (under the 450 scenario). *Source*: IEA 2015a.

needed in the medium to long term under the 2DS. Limiting the global temperature increase to 2°C requires a deep emissions reduction from current levels—more than 20 percent by 2030 (figure 1.8) and more than 50 percent by 2050. Significantly greater ambition and more actions are required by countries to generate the substantial reductions in emissions needed to meet the climate change goal. It is within this context that the shift in U.S. policy and pronouncements must be analyzed.

The change in the United States can be expected to adversely affect efforts to reduce emissions in three key distinct ways.

- First, the United States plays major leadership, political, economic, and emissions roles. It is the largest economy and one of the richest countries on a per capita basis. It is the second largest emitter after China, and it is one of the highest emitters on a per capita basis. It is also, at a global level, the premier military and international political leader. Given its combination of high levels of wealth, emissions, and influence, what the United States says and does in the arena of climate change and emissions matters a great deal. The pronouncements of the United States government questioning the impact of anthropogenic emissions on climate change and its actions to withdraw formally from the Paris Agreement have dampened the momentum created by the agreement. These actions by the United States have arguably also provided support to countries to resist taking the additional steps to achieve the significant GHG emissions reductions that the 2°C climate goal requires (set out in the "450 Scenario" depicted in figure 1.8 and in the 2DS in figure 1.1). At a time when greater ambition and more action are required, the United States shift in policy has undermined that momentum.
- Second, domestic policies do matter in promoting low-carbon options. Analysis after analysis has shown the positive impact that a supporting policy environment—and the detrimental impact of an uncertain or negative one—can have on low-carbon investments. The IEA's energy efficiency and renewables market reports of the last several years document the positive impact that supportive policies have had on these two key low-carbon technologies. The United States is taking tangible steps to eliminate a variety of policies and programs intended to support energy efficiency and renewables investments, from defunding and terminating specific initiatives to eliminating

specialized government offices to removing financial support. At the same time, the government is implementing various policies designed to promote domestic fossil fuel production. Moreover, the ongoing shift in domestic policy toward promoting "energy dominance," notably by expanding oil and gas development, exploration, and exports, will inevitably impede a shift to low-carbon technologies by the United States. Even though the country recently has enjoyed a multiyear reduction in emissions (in large part as it has shifted from coal to gas power generation), it is unlikely that the United States will be able to achieve the further and deeper longer-run emissions reductions required (by it and other nations) to achieve the 2°C goal if it does not have in place policies to promote more energy efficiency and renewables investments. Notwithstanding its recent emissions reductions, the United States remains the second largest GHG emitter. Given this context, reducing global energy GHG emissions by 50 percent by 2050 becomes increasingly difficult.

• Third, the United States is a major driver of international cooperation and financial support for climate change efforts, which directly affects other country postures and actions. A tangible illustration of the impact of the shift in U.S. policy on climate change discourse was exhibited in the 2017 G20 discussions at which the United States resisted efforts for pronouncements emphasizing the need to combat climate change as a priority. This has resulted in a split among more forward-leaning climate change countries (which previously included the United States) and less emphasis on this issue as a common global effort. This shift in U.S. policy can also be expected to influence the policies and lending patterns of multilateral development banks (MDBs) and other multilateral organizations, in particular those in which the United States is the largest shareholder. The new administration has not yet had the opportunity to put in place many of its political appointees, which has slowed its ability to implement in practice its shift in policies; this is particularly true for areas where it wants to implement a dramatic move away from the policies of the eight-year Obama administration, as is the case for climate change. So, for example, while many MDBs have recently prioritized lending for low-carbon investments, a weakening of these efforts over the next several years would not be surprising as newly appointed United States officials assume their positions and make their preferences known within these organizations. International finance is particularly important to fund low-carbon investments in developing countries whose emissions are, absent these types of interventions, projected to substantially increase as their populations and income levels increase; a reduced willingness to support these countries in making the transition to a low-carbon development pathway, notably through targeted financial assistance, will weaken efforts to reduce global emissions.

As a consequence of these factors, the effort to reduce emissions dramatically as is needed to limit the global temperature increase to 2°C or less, which was already a difficult task, has now become even more so. United States action and leadership are important to achieve the goals of the Paris Agreement. While other countries (such as China and various European nations) continue to promote the climate change agenda, the dramatic shift in U.S. policy in this regard will inevitably weaken global efforts to reduce anthropogenic GHG emissions. And so, we can expect more CO_2 in our atmosphere and we can anticipate a hotter planet in our future.

Notes

1. See discussion in chapter 1 of *Energy, Climate Change and Environment: 2016 Insights* (IEA 2016a). This publication is referred to hereinafter as "ECCE 2016".
2. Set out in Article 2 of the Paris Agreement. The clause also encourages parties to "pursue efforts to limit the temperature increase to 1.5°C" . . . in all cases relative to preindustrial levels.
3. See additional discussions in ECCE 2016, as well as in *20 Years of CCS* (IEA 2016d).
4. See, for example, the discussion of complementing moderate carbon prices with other policies to push the energy transition in ECCE 2016 (IEA 2016a).
5. See fuller discussion in chapter 2 of ECCE 2016 (IEA 2016a). Based on work drafted by Matthew Gray and Philippe Benoit.
6. See fuller discussion in chapter 5 of ECCE 2016 (IEA 2016a). Based on work drafted by Philippe Benoit and David Morgado.
7. This positive impact can be limited by the rebound effect, in which energy efficiency benefits encourage some increase in energy consumption which then limits anticipated energy savings. See discussion of rebound effect in *Capturing the Multiple Benefits of Energy Efficiency* (IEA 2015c).

8. Although energy efficiency is often expressed from an end-user's perspective, i.e., impact on TFC, it is primary energy demand that is a more relevant measure for climate change purposes since it covers the total fossil fuel combustion (and corresponding GHG emissions generated) to meet that demand.
9. This is distinguished from the various energy conservation programs implemented in the face of a supply crisis that are designed to reduce consumption or otherwise modify consumer behavior, even with a resulting potentially lower standard of living over the short term in order to face the energy supply crisis.
10. This program is described at https://www.iea.org/media/workshops /2015/eeuevents/behave1103/S7LindaCifolelliEnea.pdf.
11. See presentation at https://www.youtube.com/watch?v=fJxpyqN1SPg.
12. See fuller discussion in chapter 6 of ECCE 2016 (IEA 2016a). Based on work drafted by Philippe Benoit, Liwayway Adkins, and George Kamiya.
13. The IEA has estimated that over 90 percent of future energy demand growth will occur under a business-as-usual case in these countries (IEA 2014b).
14. For capacity ownership estimates in this section, state ownership is defined as majority owned or controlled by governments.
15. SOEs own about one-third of the CCS projects under operation or under construction; see *20 Years of CCS*, section 1.5 (IEA 2016d).

References

Benoit, P. 2012. "State-Owned Enterprises and Their Domestic Financial Base: Two Keys to Financing Our Low-Carbon Future." In *Electricity in a Climate-Constrained World*, edited by R. Baron and J. Lampreia, 25–32. Paris: OECD/IEA.

Earnhart, D.H., M. Khanna, and T.P. Lyon. 2014. "Corporate Environmental Strategies in Emerging Economies." *Review of Environmental Economics and Policy* 8 (2): 164–185.

IEA (International Energy Agency). 2016a. *Energy, Climate Change and Environment: 2016 Insights.* Paris: OECD/IEA.

IEA. 2016b. *Energy Technology Perspectives 2016.* Paris: OECD/IEA.

IEA. 2016c. *Energy Efficiency Market Report 2016.* Paris: OECD/IEA.

IEA. 2016d. *20 Years of CCS.* Paris: OECD/IEA.

IEA. 2016e. *World Energy Investment 2016.* Paris: OECD/IEA.

IEA. 2015a. *Energy and Climate Change: World Energy Outlook Special Report.* Paris: OECD/IEA.

IEA. 2015b. *World Energy Outlook 2015*. Paris: OECD/IEA.

IEA. 2015c. *Capturing the Multiple Benefits of Energy Efficiency*. Paris: OECD/IEA.

IEA. 2015d. *CO_2 Emissions from Fuel Combustion*. Paris: OECD/IEA.

IEA. 2014a. *Energy Technology Perspectives 2014*. Paris: OECD/IEA.

IEA. 2014b. *World Energy Outlook 2014*. Paris: OECD/IEA.

IEA. 2011. *Saving Electricity in a Hurry*. Paris: OECD/IEA.

OECD (Organisation for Economic Co-operation and Development). 2015a. *State-Owned Enterprises in the Development Process*. Paris: OECD Publishing.

OECD. 2015b. *OECD Environmental Performance Reviews: Brazil 2015*. Paris: OECD Publishing.

OECD. 2014. *The Size and Sectoral Distribution of SOEs in OECD and Partner Countries*. Paris: OECD Publishing.

CHAPTER TWO

Transitional Risks and the Safe Carbon Budget

RICK VAN DER PLOEG

Introduction

Various economic studies derive optimal climate policies from maximizing social welfare subject to the constraints of an integrated assessment model that combines both a model of the economy and a model of the carbon cycle and temperature dynamics (e.g., Nordhaus 1991, 2010, 2014; Golosov, Hassler, Krusell, and Tsyvinski 2014; Dietz and Stern 2015; Rezai and van der Ploeg 2016). The resulting carbon price is more or less proportional to world gross domestic product (GDP) if global warming damages are proportional to GDP. This price depends on ethical considerations such as intergenerational inequality aversion (the lack of willingness to sacrifice consumption today to curb global warming many decades into the future) and impatience or the amount by which welfare of future generations is discounted. It also depends on detailed aspects of the carbon cycle and heat exchange dynamics (e.g., the fraction of carbon emissions that stays up permanently and the rate at which the remaining parts of the carbon stock returns to the surface of the earth).

In contrast, the Intergovernmental Panel on Climate Change (IPCC) eschews a welfare-maximizing approach and instead focuses on a clear and easy-to-communicate target for peak global warming of 2°C or perhaps 1.5°C and specifies a probability of two-thirds that this target must be met, which corresponds to a risk tolerance of a one-third. Since cumulative carbon emissions drive peak global warming, the target for peak global warming determines how much more carbon can be emitted in total. This will be called the *safe carbon budget* and depends on three key parameters: maximum permissible global warming, climate uncertainty, and risk tolerance.

Corresponding to the safe carbon budget there is a constrained efficient (cost-minimizing) time path for the price of carbon which ensures that cumulative emissions from now on stay within the safe carbon budget. For our purposes, this adjusted optimal carbon price is not the cost-minimizing Hotelling-type price calculated in previous studies on temperature constraints (e.g., Nordhaus 1982; Tol 2013; Lemoine and Rudik 2017), but the welfare-maximizing carbon price that follows from adjusting the production damages from global warming upward in such a way that the safe carbon budget constraint is never violated. This carbon price can, together with the time paths for mitigation and abatement, be derived from an integrated assessment model and is higher than the unadjusted optimal carbon price; it also rises at the same rate as GDP. As a result, one can determine how fast fossil fuel is phased out and renewable energies are phased in and how much fossil fuel is abated. Using the concept of the safe carbon budget means that ethical concepts such as how much to discount the welfare of future generations and the willingness to sacrifice consumption today to curb global warming play no role in determining the safe budget, but do affect the timing of the energy transition and how much fossil fuel is abated.

1. Paris COP21 Target for Peak Global Warming and the Safe Carbon Budget

The key driver of peak global warming measured as deviation from preindustrial temperature, *PGW*, is cumulative carbon emissions, *E* (Allen et al. 2009; IPCC 2013; Allen 2016). Denoting the transient climate response by *TCR*, a simple reduced-form relationship is:

$$PGW = \alpha + TCR \times E \quad \text{with} \quad TCR \equiv \overline{TCR} \times \varepsilon$$
$$\text{and} \quad \ln(\varepsilon) \sim N(\mu, \sigma^2), \tag{2.1}$$

where α is a constant, $\overline{TCR}$ is the mean TCR, ε is a lognormally distributed shock to the TCR with the mean set to $\mu = -0.5\sigma^2$ so that $E[\varepsilon] = 1$. This formulation allows us to investigate the effects of mean-preserving spreads in the TCR by simply varying σ. Uncertainty in the TCR may follow from a more complicated stochastic process with dynamics and non-normal features such as fat tails and may result from a number of underlying shocks to the climate system, but equation (2.1) keeps it simple. Paris COP21 has agreed to keep PGW below 2°C (and aim at 1.5°C). Let us assume that this target has to be met with the probability $0 < \beta < 1$:

$$\text{prob}[PGW < 2°C] = \beta. \qquad (2.2)$$

IPCC typically sets β to 2/3. The safe carbon budget compatible with equation (2.2) is from equation (2.1):

$$E \leq \frac{2-\alpha}{\overline{TCR} \times \exp\left(F^{-1}(\beta; -0.5\sigma^2, \sigma^2)\right)} \equiv \overline{E}, \qquad (2.3)$$

where $F(.; \mu, \sigma^2)$ is the cumulative normal density function with mean μ and variance σ^2. Equation (2.3) indicates that a more ambitious target for peak global warming, say 1.5°C, instead of 2°C, a higher expected transient climate response to cumulative emissions, or a lower-risk tolerance (higher value of β) imply that less carbon can be burned and thus more fossil fuel must thus be locked up in the earth. Furthermore, uncertainty about the transient climate response keeping the expected TCR constant (higher σ^2) also cuts the maximum tolerated emissions or the safe carbon budget.

Without uncertainty, a carbon budget of $E = (PGW - \alpha) / \overline{TCR} =$ 362 GtC is compatible with PGW of 2°C if $\alpha = 1.276$°C and the transient climate response is 2°C per trillion tons of carbon (Allen 2016; van der Ploeg and Rezai 2016). McGlade and Ekins (2015) show that reserves and probable reserves (resources) are a factor 3 to 10–11 times higher than the carbon budget compatible with peak temperatures of 2°C. They calculate that 80 percent of coal reserves, half of gas reserves, and a third of oil reserves must be left unburned. In practice, much more may need to be abandoned as many oil and gas reserves are owned by states instead of private companies. Not only

TABLE 2.1
Risk Tolerance and the Safe Carbon Budget (GtC)

Risk Tolerance = $1 - \beta$	1/3		10 Percent		1 Percent	
Standard deviation of lognormal shock to $TCR = \sigma$	0.2	0.6	0.2	0.6	0.2	0.6
Safe carbon budget if $PGW = 2°C$	339	335	281	201	232	107
Safe carbon budget if $PGW = 1.5°C$	105	104	87	62	72	33

Note: $\alpha = 1.276°C$ and expected = 2°C/TtC.

will carbon assets be stranded but also energy-intensive irreversible investments in electricity generations such as coal-fired stations.

Equation (2.3) indicates that allowing for climate risk implies a lower safe carbon budget and more stranded assets. Table 2.1 illustrates this with some illustrative mean-preserving increases in uncertainty in the *TCR* for peak global warming targets of both 2°C and 1.5°C. Focusing at a *PGW* target of 2°C, table 2.1 indicates that a risk tolerance of 1/3 (in line with the IPCC) gives a safe carbon budget of 339 GtC if $\sigma = 0.2$ and 235 GtC if $\sigma = 0.6$. Tightening up risk tolerance to 10 and 1 percent curbs the safe carbon budget to 281 GtC and 232 GtC if $\sigma = 0.2$ and 201 and 107 GtC if $\sigma = 0.6$. More uncertainty in the *TCR* thus implies that less carbon can be burned in total.

If *PGW* has to be kept below 1.5°C, the safe carbon budget without uncertainty drops dramatically from 362 GtC to 112 GtC. If $\sigma = 0.6$, the safe carbon budget drops to 104, 62, or 33 GtC depending on whether the risk tolerance is 10, 1, or 0.1 percent, respectively.

2. Optimal Energy Transition Given the Safe Carbon Budget

What is the optimal timing of fossil fuel use and carbon emissions, the mitigation and abatement rates, and the end of the fossil fuel era? These depend crucially on the costs of fossil fuel including any price charged for carbon emissions versus that of renewable energy, the cost of abatement, and the various rates of technical progress. It is thus not surprising that the Paris Agreement stresses a tight target for *PGW* with reference to geophysical conditions and risk while economists highlight the intertemporal costs and benefits of global warming, ethical attitudes to intertemporal trade-offs, risk aversion, and prudence. To illustrate this, I augment a very simple integrated assessment model put forward in van der Ploeg and Rezai (2016) with the constraint on the safe carbon budget (2.3). This model has a constant trend growth in world GDP, g, and a constant rate of technological progress in fossil fuel extraction, mitigation of energy (i.e., raising the share of renewable energy), and abatement. It has a two-box carbon cycle (cf. Golosov et al. 2014) and a simple lag between temperature and increases in atmospheric carbon concentration. Details of the calibration are omitted, but they are based on Integrated Assessment Model of Climate and the Economy, DICE (Nordhaus 2010, 2014).

Maximizing global welfare subject to the resource constraint that income available after damages has to equal spending on consumption, energy generation, mitigation, and abatement yields the *unadjusted* optimal carbon price, and maximizing the welfare subject to the additional constraint that cumulative carbon emissions cannot exceed the safe carbon budget yields the *adjusted* optimal carbon price. The unadjusted optimal price of carbon requires additional parameters for the carbon cycle, i.e., the fraction of carbon emissions staying up in the atmosphere forever, β_0, the rate of return of remaining emissions to the surface of the earth and oceans, β_1, and the mean lag between the temperature rise following an increase in atmospheric carbon, $Tlag$, and for the ethical considerations, i.e., the rate at which the welfare of future generations is discounted, RTI, and the coefficient of relative intergenerational inequality aversion, IIA. The unadjusted optimal carbon price is then equal to:

$$P_t = \tau^U \times WGDP_t \text{ with } \tau^U$$

$$\equiv \left(\frac{\beta_0}{SDR} + \frac{1-\beta_0}{SDR + \beta_1}\right)\left(\frac{1}{1 + SDR \times Tlag}\right)d, \qquad (2.4)$$

where $WGDP_t$ denotes world GDP at time t and $SDR \equiv RTI + (IIA - 1) \times g$ is the growth-corrected social discount rate. The unconstrained optimal carbon tax is thus high and climate policy ambitious if a large part of emissions stay up forever (high β_0), the absorption rate of the oceans is low (low β_1), the temperature lag is small, the welfare of future generations is discounted less heavily (low RTI), and there is less willingness to sacrifice consumption to curb future global warming (low *IIA*). With higher economic growth (high *g*) future generations are richer so current generations are less prepared to curb global warming, but growth in damages from global warming is also higher and thus a higher carbon price is warranted. The net effect of higher growth on the unadjusted optimal carbon price is negative provided that *IIA* > 1.

If the safe carbon budget constraint in equation (2.3) bites, the adjusted optimal carbon price, P_t, results:

$$P_t = \tau^C \times WGDP_t \text{ with } \tau^C > \tau^U \text{ from } E(\bar{t})$$

$$= \int_0^{\bar{t}} (1 - a_t)(1 - m_t)\gamma_0 e^{-r_\gamma t} \Upsilon_0 e^{gt} = \bar{E}, \tag{2.4'}$$

where $m(t)$ is the mitigation rate (the share of renewables in total energy) at time t, $a(t)$ the abatement rate at time t, $\gamma_0 e^{-r_\gamma t}$ is energy use as a fraction of world GDP at time t, and $\bar{t}$ is the date of the end of the fossil fuel era. The adjusted optimal price of carbon (2.4′) rises with trend growth. It does not depend on damages from higher temperatures, how patient society is, how much intergenerational inequality aversion there is, or on the geophysical parameters such as the absorption capacity of the oceans or the temperature lag. However, the unadjusted optimal carbon price (2.4) does depend on these parameters.

Cost minimization given the carbon price (equations 2.4 or 2.4′) requires that the marginal cost of extracting fossil fuel equal the marginal cost of mitigating fossil fuel plus the price of carbon for using unabated fossil fuel, $(1 - a_t)P_t$. Mitigation thus increases the relative cost of carbon-emitting technologies and abatement including the price of nonabated carbon. Cost minimization requires that the marginal cost of abatement equal the saved cost of carbon emissions. Abatement thus rises as its cost falls or the price of carbon rises over time. I assume cost conditions are such that fossil fuel is fully mitigated before it is fully abated.

3. Optimal Climate Policy Simulations with a Safe Carbon Budget

Table 2.2 gives our benchmark estimates of the variance of the log-normally distributed shock to the *TCR*, the target for *PGW*, and risk tolerance, and some of the additional parameters needed for calculation of the unconstrained optimal climate policy.

TABLE 2.2
Calibration Details

Mean transient climate response to cumulative emissions	$TCR = 2°C/TtC$, $\alpha = 1.276°C$
Variance of the lognormal shock to the $TCRE$	$\sigma = 0.6$
Target for peak global warming	$2°C$
Risk tolerance	$1 - \beta = 1/3$
Growth rate in world GDP	$g = 2$ percent per year
Additional parameters needed for the unadjusted optimal climate policy	
Rate of time patience	$RTI = 1.5$ percent per annum
Intergenerational inequality aversion	$IIA = 1.45$, Relative risk aversion = $RRA = 1.45$
Growth-corrected social discount rate	$SDR = RTI + (IIA - 1) \times g = 2.45$ percent per year
Flow damage of global warming	$d = 1.9$ percent of GDP per TtC

Using this calibration, not pricing carbon at all leads to zero mitigation and zero abatement, cumulative emissions of 1778 GtC, 118 years for the end of the fossil fuel era to occur, and *PGW* of 4.6°C, which is much too high. The globally best *unadjusted* optimal climate policy summarized by the *solid* lines in figure 2.1 shows an initial price of carbon is \$44/tC (or \$12/tCO$_2$), and grows at 2 percent per annum from then on. The mitigation rate is driven by technological progress and the rising price of carbon, and rises from 20 to 100 percent in seventy-eight years at which date the carbon-free era starts. The abatement rate rises from a mere 1.5 to 19 percent at the end of the fossil fuel era. Therefore, 635 GtC is burned, which implies a *PGW* of 2.6°C.

The unadjusted climate policy thus overshoots the 2°C target agreed at the Paris COP21 conference. The *dashed* lines in figure 2.1 therefore show the benchmark *adjusted* optimal time paths of the price of carbon, the mitigation ratio, and the abatement ratio corresponding to a constraint on the safe carbon budget of 362 GtC corresponding to a *PGW* target of 2°C and no uncertainty in the *TCR*.

The end of the fossil fuel era then occurs more quickly: after sixty-three instead of seventy-eight years. The price of carbon starts at more than double, i.e., at \$92/tC (or \$25/tCO$_2$), and then rises

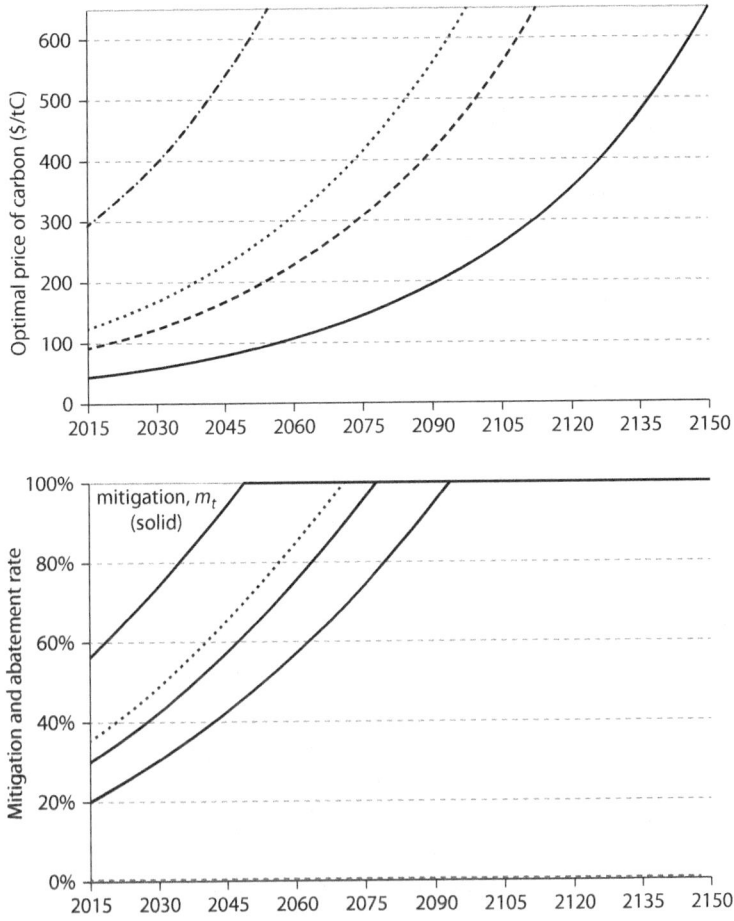

Figure 2.1 Constrained and unconstrained optimal climate policy.

Note: Solid lines: unconstrained optimal climate policy and cumulative emissions of 635 GtC. *Dashed lines:* constrained optimal climate policy with 2°C target and no uncertainty about *TCRE* corresponds to a safe carbon budget of 362 GtC. *Dotted lines:* constrained optimal climate policy with 2°C target, uncertainty about the *TCRE* and risk tolerance of 10% corresponds to safe carbon budget of 271 GtC. *Dashed-dotted lines:* constrained optimal climate policy under uncertainty and target of 1.5°C corresponds to a safe carbon budget of 84 GtC and a much more ambitious climate policy.

Mitigation under business as usual (there is no abatement under business as usual) gives cumulative emissions of 1,778 GtC.

at 2 percent per annum in line with the rate of economic growth. As a result of this more ambitious climate policy, the path for the mitigation rate is higher and starts at 30 percent and rises to 100 percent during the fossil fuel era. The abatement rate is also higher; it starts at 3.1 percent and rises to 26.9 percent toward the end of the fossil fuel era. This adjusted optimal climate policy corresponds to raising the flow damages of higher stocks of carbon in the atmosphere from 1.9 to 4 percent of GDP per TtC.

The *dotted* lines in figure 2.1 show how these time paths for the adjusted optimal climate policy are affected by uncertainty in the *TCR*. With $\beta = 10$ percent and $\sigma = 0.2$, the safe carbon budget is 281 GtC. The transition to the carbon-free era occurs earlier driven by a higher path for the carbon price, which starts at \$121/tC (\$33/tCO$_2$) and grows at a rate of 2 percent per year. As a result, the time paths for the mitigation rate and the abatement rate are lifted up. This ensures that in total, less carbon is burned and that the expected *PGW* is less than 2°C to be on the safe side.

This corresponds to boosting the flow damages from 1.9 or 4.0 percent to 5.3 percent of GDP per TtC. An even smaller risk tolerance or higher variance of the shock to the transient climate response would lead to a smaller safe carbon budget and quicker transition to the carbon-free era and higher mitigation and abatement rates, driven by a higher price of carbon throughout.

The dashed lines for the adjusted optimal carbon price will be lifted up if climate uncertainty is higher or risk tolerance is tightened up as then the safe carbon budget is smaller (see table 2.1).

4. Conclusion

The Paris COP21 target of 2°C or 1.5°C implies a safe carbon budget of 362 or 112 GtC, respectively. Climate uncertainty and a tighter risk tolerance imply that even less fossil fuel can be burned in total, thus requiring a more ambitious climate policy. Since damages from global warming are relatively modest, the 2°C and safe carbon budget constraints bite. These targets are easier to negotiate and communicate, and do not depend on various ethical considerations regarding the welfare of current and future generations. The safe carbon budget constraint effectively almost triples the damage flow per ton of carbon in the atmosphere.

Uncertainty about growth of aggregate consumption depresses the social discount rate of a prudent policy maker and pushes up the optimal price of carbon even more (e.g., Gollier 2012). Other types of uncertainty such as uncertainty about the damage flows resulting from atmospheric carbon, climate sensitivity, and the sudden release of greenhouse gases into the atmosphere boost the optimal price of carbon even further. The risks of interacting, multiple tipping points lead to a two- to eightfold increase in the carbon price (Cai, Lenton, and Lontzek 2016; Lemoine and Traeger 2016).

The 2°C target for peak global warming with a risk tolerance of 10 percent and a modest degree of climate uncertainty gives a climate budget of 281 GtC. To ensure cumulative emissions remain with this budget, a carbon price achieved via a global tax or a global competitive emissions market, of $121 per ton of carbon (or $33 per ton of CO_2) that grows at a rate of 2 percent per year is required. Mitigating risks of multiple tipping points may lead to a carbon price factor three times higher. Sadly, the likelihood of this happening is small as countries are unlikely to agree on the international transfers from rich to poor countries that are needed to sustain such a global carbon price. Although the nonclimate benefits of mitigating global warming gives countries a rational individual incentive to price carbon, climate policy will be much too lackluster if an international climate deal is not forthcoming.

Notes

Presented at the conference "The Energy Transition, NDCs, and the Post-COP 21 Agenda," Marrakesh, September 8–9, 2016, organized by the COP22-Marrakesh IMF and OCP at SIPA, Columbia University. I am very grateful to Armon Rezai for our collaborations on the integrated assessment model I use here and for Ton van den Bremer's helpful comments.

References

Allen, M. R. et al. 2009. "Warming Caused by Cumulative Emissions Towards the Trillionth Tonne." *Nature* 458: 1163–1166.

Allen, M. 2016. "Drivers of Peak Warming in a Consumption-Maximizing World." *Nature Climate Change*, no. 6, 684–686.

Cai, Y., T. M. Lenton, and T. S. Lontzek. 2016. "Risk of Multiple Climate Tipping Points Should Trigger a Rapid Reduction in CO_2 Emissions." *Nature Climate Change* 6, 520–525.

Dietz, S., and N. Stern. 2015. "Endogenous Growth, Convexity of Damages and Climate Risk: How Nordhaus' Framework Supports Deep Cuts in Emissions." *Economic Journal* 125 (583): 574–620.

Gollier, C. 2012. *Pricing the Planet's Future: The Economics of Discounting in an Uncertain World*. Princeton, New Jersey: Princeton University Press.

Golosov, M., J. Hassler, P. Krusell, and A. Tsyvinski. 2014. "Optimal Taxes on Fossil Fuel in General Equilibrium." *Econometrica* 82 (1): 48–88.

IPCC (Intergovernmental Panel on Climate Change). 2013. "Long-Term Climate Change: Projections, Commitments, and Irreversibilities." Working Group 1, Contribution to the *IPCC 5th Assessment Report*. International Panel of Climate Change, chapter 12, sections 5.4.2 and 5.4.3.

Lemoine, D., and I. Rudik. 2017. "Steering the Climate System: Using Inertia to Lower the Cost of Policy." *American Economic Review* 107 (10): 2947–2957.

Lemoine, D., and C. P. Traeger. 2016. "Economics of Tipping the Climate Dominoes." *Nature Climate Change*, no. 6, 514–519.

McGlade, C., and P. Ekins. 2015. "The Geographical Distribution of Fossil Fuels Unused When Limiting Global Warming to 2°C." *Nature*, no. 517, 187–190.

Nordhaus, W. 1982. "How Fast Should We Graze the Global Commons?" *American Economic Review* 72 (2): 242–246.

Nordhaus, W. 1991. "To Slow or Not to Slow: The Economics of the Greenhouse Effect." *Economic Journal* 101 (407): 920–937.

Nordhaus, W. 2010. "Economic Aspects of Global Warming in a Post-Copenhagen World." *Proceedings of the National Academy of Sciences* 107 (26): 11721–11726.

Nordhaus W. 2014. "Estimates of the Social Cost of Carbon: Concepts and Results from the DICE-2013R Model and Alternative Approaches." *Journal of the Association of Environmental and Resource Economists* 1 (1/2): 273–312.

Rezai, A., and F. van der Ploeg. 2016. "Intergenerational Inequality Aversion, Growth and the Role of Damages: Occam's Rule for the Global Carbon Tax." *Journal of the Association of Environmental and Resource Economists* 3 (2): 493–522.

Tol, R.S.J. (2013). "Targets for Global Climate Policy: An Overview." *Journal of Economic Dynamics and Control* 37 (5): 911–928.

van der Ploeg, F., and A. Rezai. 2016. "Climate Policy with Declining Discount Rates in a Multi-Region World—Back-on-the-Envelope Calculations" mimeo.

PART TWO

Carbon Pricing and Dealing with Uncertainty

CHAPTER THREE

Fighting Climate Change and the Social Cost of Carbon

CHRISTIAN GOLLIER

1. This Is the Tragedy of the Commons!

Climate change is expected to dramatically deteriorate the well-being of future generations, and the Paris Accord of December 2015 has not solved the problem. Although the precise consequences of our inaction are still hard to quantify, there is no question that a business-as-usual scenario would be catastrophic. The Fifth Assessment Report of the Intergovernmental Panel on Climate Change (IPCC 2014) estimates that the average temperature will increase by somewhere between 2.5°C and 7.8°C by the end of this century, after having already increased by almost 1°C over the last century. Despite the emergence over the last three decades of solid scientific information about the climate impacts of increased CO_2 concentration in the atmosphere, the world's emissions of greenhouse gas (GHG) have never been larger, rising from 30 $GtCO_2eq$/year in 1970 to 49 $GtCO_2eq$/year in 2010. According to the IPCC, about half of the anthropogenic CO_2 emissions between 1750 and 2010 occurred during the last four decades (figure 3.1), mainly as a result of economic and population growth and the dearth of actions to fight climate change.

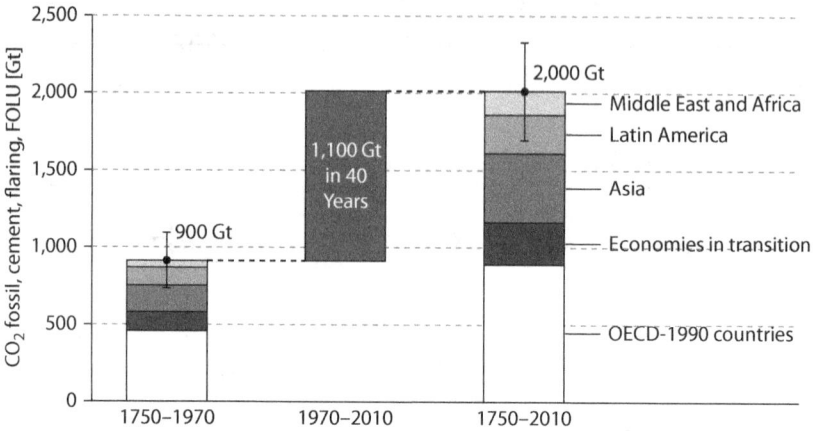

Figure 3.1 Emissions of CO_2 since 1750. *Source*: IPCC (2014).

Limiting the increase in temperature to 2°C is thus an immense challenge, with a still increasing world population and, hopefully, more countries accessing Western standards of living. It will require radical transformations in the way we use energy, including the way we heat and locate our houses, transport people, and produce goods and services.

However, most benefits of mitigation are *global* and *distant*, while costs are local and immediate. The geographic and temporal dimensions of the climate problem account for the current inaction. Climate change is a global commons problem. In the long run, most countries, states, and individuals will benefit from a massive reduction in global emissions of GHGs, but individual incentives to do so are negligible. Most of the benefits of a country's efforts to reduce emissions go to other countries. In a nutshell, a country bears 100 percent of the cost of a green policy and receives, say, 1 percent of the benefits of the policy if the country has 1 percent of the population and has an average exposure to climate-related damages. Consequently, countries do not internalize the benefits of their mitigation strategies, emissions are high, and climate changes dramatically. The free-rider problem is well-known to generate the "tragedy of commons" (Hardin 1968). A country or region that contemplates a unilateral mitigation strategy is further discouraged by the presence of so-called "carbon leakages." Namely, imposing additional costs to high-emission domestic industries makes them noncompetitive. This tends to move production to less responsible countries, yielding an international redistribution of production and wealth with negligible ecological benefit. Similarly, the reduction in demand for fossil energy originating from the virtuous countries tends to reduce their international price, thereby increasing the demand and emissions in nonvirtuous countries. This other carbon leakage also reduces the net climate benefit of the effort made by any incomplete club of virtuous countries. Its intertemporal version is called the *green paradox*. It states that a commitment to be green in the future leads oil producers to increase their production to cater to today's nonvirtuous consumers. Since carbon sequestration is not a mature technology, mitigation is a threat to the oil rent, and its owners should be expected to react to this threat.

One of the most difficult challenges of climate change comes from the existence of a large fossil fuels rent currently owned by resource-rich countries. This rent exists because of the relative scarcity of the reserves of these nonrenewable resources, and the expectation of a future exhaustion or at least steeply increasing marginal costs of extraction. The problem is that these reserves are large, as shown in figure 3.2. The cumulated consumption (middle portion of bars in the graph) of gas, coal, and oil since the beginning of the Industrial Revolution has been quite limited compared to the stock of these

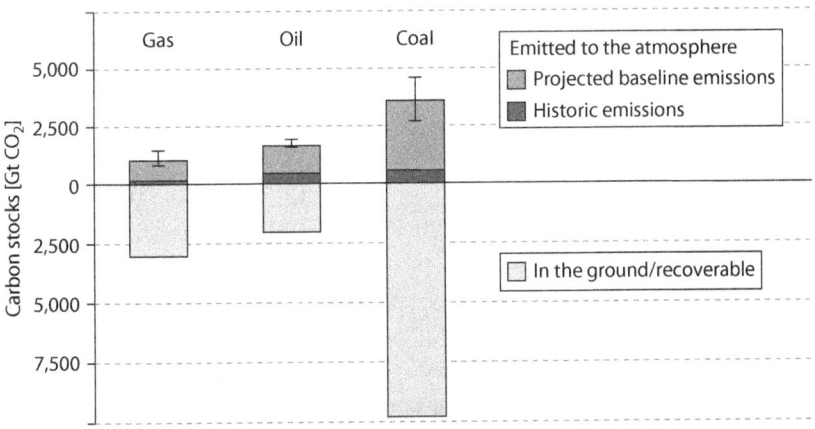

Figure 3.2 Past consumption and current reserves of fossil fuels. *Source*: IPCC (2014).

resources. Adding consumption until the end of this century (top portion of bars) in the business-as-usual scenario will still leave most of the stock in the ground. The burning of the entire stock of fossil resources on this planet within the next two centuries or so would certainly devastate our planet by raising GHG concentration way above the acceptable limits. If an efficient and credible climate policy were to be implemented, this would imply the annihilation of the fossil fuels rent, and a flood of stranded assets in carbon-intensive sectors around the world. Its strategic and geopolitical consequences shed some light on the difficulty in reaching an international agreement involving oil-rich countries.

The good news is that an efficient international climate agreement will generate an important social surplus to be shared among the world's citizens. The political economy of climate change, however, is unfavorable: The costs of any such agreement are immediate, whereas most benefits will occur in the distant future, mainly to people who are not yet born and *a fortiori* do not vote. In short, climate mitigation is a long-term investment. Many activists and politicians promote climate mitigation policies as an opportunity to boost "economic growth." The fact that no country (with the exception of Sweden) comes remotely close to doing its share should speak volumes: Why would countries sacrifice the consumption of goods and leisure to be environmentally unfriendly? The reality is bleaker, in particular for economies in crisis and in the developing world. In reality, fighting climate change implies reducing consumption in the short run to finance green investments that will generate a better environment only in the distant future. It diverts economic growth from consumption to investment, not good news for the well-being of the current poor. Carbon pricing, if implemented, will induce households to invest in photovoltaic panels on their roofs or to purchase expensive electric cars, actions that yield no obvious increase in their own well-being, to the detriment of spending the corresponding income on other goods.

To be certain, countries may perceive some limited *co-benefits* of climate-friendly policies. For example, green choices may also reduce emissions of other pollutants (coal plants produce both CO_2 and SO_2, a regional pollutant); in a similar spirit, countries may encourage their residents to eat less red meat not so much from a global warming concern but because they want to reduce the occurrence of cardiovascular diseases. Replacing dirty lignite by gas and oil as the main source of energy had enormous sanitary and environmental benefits in Western countries after World War II (WWII), for example, by eliminating smog from London. Therefore, *some* actions are to be expected from countries with an eye on national interests only (not to mention the political benefits of placating domestic and international opinion). But these "zero ambition" actions will be far insufficient to generate what it takes to keep global warming manageable.

Overall, fighting climate change yields short-term collective costs, thereby creating a political problem for benevolent decision makers who support an ambitious international agreement. In sum, without a collective incentive mechanism, one's investment in a responsible mode of living will hardly benefit one's well-being. Rather, and assuming away leakages, it will benefit distant generations who will mostly live in other countries. It is collectively efficient to act, but individually optimal to do little.

2. A Uniform Carbon Price Is Necessary

The core of the climate externality problem is that economic agents do not internalize the damages that they impose on other economic agents when they emit GHGs. The approach that economists have long proposed to solve the free-rider problem consists of inducing economic agents to internalize the negative externalities that they impose when they emit CO_2 (*polluter pays principle*). This is done by pricing at a level corresponding to the present value of the marginal damage associated to the emission, and by forcing all emitters to pay this price. Because GHGs generate the same marginal damage regardless of the identity of the emitter and of the nature and location of the activity, all tons of CO_2 should be priced equally. By imposing the same price on all economic agents around the world, it would ensure that all actions to abate emissions that cost less than that price will be implemented. This least-cost approach guarantees that the reduction of emissions necessary to attain the global concentration objective will be made at the minimum global cost.

In contrast with this economic approach, *command-and-control* approaches (source-specific emissions limits, standards and technological requirements, uniform reductions, subsidies/taxes that are not based on actual pollution, vintage-differentiated regulations, industrial policy, etc.) usually create wide discrepancies in the implicit price of carbon placed on different emissions. This has been shown empirically to lead to substantial increases in the cost of environmental policies.

Western countries have made some attempts at reducing GHG emissions, notably through direct subsidization of green technologies: generous feed-in electricity tariffs for solar and wind energy, bonus-malus systems favoring low-emission cars, subsidies to the biofuel industry, etc. For each green policy one can estimate its implicit carbon price, i.e., the social cost of the policy per ton of CO_2 saved. A recent Organization for Economic Co-operation and Development study (OECD 2013) showed that these implicit prices vary widely across countries, and also across sectors within each country. In the electricity sector, OECD estimates range from less than zero to $900. In the road transportation sector, the implicit carbon price can be as large as $1000, in particular for biofuels. The high heterogeneity of implicit carbon prices in actual policy making is a clear demonstration of the inefficiency of this command-and-control approach. Similarly, any global agreement that does not include all world regions in the climate coalition will exhibit the same inefficiency by setting a zero-carbon price in nonparticipating countries.

While economists for good reasons are broadly suspicious of command-and-control policies, they also understand that these policies may occasionally be a second-best solution when measurement or informational problems make direct pricing complex and/or when consumers discount the future too much. This is the classic justification for housing insulation standards, for instance. But command-and-control approaches are best avoided when feasible.

Income inequality is certainly an important issue, but its solution should not be found in a Kyoto Protocol–like manipulation of the law of a single carbon price. The non-Annex 1 parties of the Kyoto Treaty had no binding obligation and their citizens faced no carbon price. This derailed the ratification of the protocol by the U.S. Senate. The Clean Development Mechanism (CDM) designed in Kyoto was aimed at alleviating the imperfect coverage problem; it met with limited success and was also not a satisfactory approach due to yet another leakage problem. For example, Annex 1 countries paying to protect a forest in a less developed country increase the price of whatever the deforestation would have been allowed to sell (beef, soy, palm, or wood) and encourages deforestation elsewhere. The CDM mechanism also created the perverse incentive to build, or maintain in operation longer than planned, polluting plants in order to later claim CO_2 credits for their reduction.

The best way to implement a uniform price of carbon is the so-called *cap-and-trade strategy*. Under this solution, the agreement would specify a worldwide, predetermined number (the cap) of tradable emissions permits. The tradability of these permits will ensure that countries face the same carbon price, emerging from mutually advantageous trades on the market for permits; the cross-country price here would not result from an agreed-upon price of carbon, but rather from a clearing in this market. To address compensation, permits would be initially allocated to the different countries or regions, with an eye on getting all countries on board (redistribution). But in the long run, all economic agents will have to pay the market price of carbon for the right to emit greenhouse gases.

The cap-and-trade system was adopted, albeit with a failed design, by the Kyoto Protocol. The Kyoto Protocol of 1997 extended the 1992 United Nations Framework Convention on Climate Change (UNFCCC) that committed participating countries to reduce their emissions of GHG. The treaty entered into effect on February 16, 2005. The Annex B parties committed to reduce their emissions in 2012 by 5 percent compared to 1990, and to use a cap-and-trade system. Kyoto participants initially covered more than 65 percent of global emissions in 1992. But the nonratification by the United States and the withdrawal of Canada, Russia, and Japan, combined with the boost of emerging countries emissions reduced the coverage to less than 15 percent in 2012. The main real attempt to implement a carbon-pricing mechanism within the Kyoto agreement emerged in Europe, with the European Union Emission Trading Scheme (EU ETS). In its first trading period of 2005–2007 ("phase 1"), the system was established with a number of allowances (the so-called Assigned Amount Units, AAUs) based on the estimated needs; its design was flawed in many respects, and far inferior to what the United States adopted in 1990 to reduce SO_2 emissions by half. In the second trading period of 2008–2012, the number of allowances was reduced by 12 percent in order to reduce the emissions of the industrial and electricity sectors of the European Union. This crackdown was offset by the possibility given to the capped entities to use the Kyoto offsets (mostly from the Clean Development Mechanism) for their compliance.

In addition, the deep economic crisis that hit the region during the period reduced the demand for permits. Moreover, large subsidies in the renewable energy sector implemented independently in most countries of the European Union further reduced the demand for permits. In the absence of any countervailing reaction on the supply of permits, the carbon price went down from a peak of €30/tCO$_2$ to around €5–7/tCO$_2$ today. This recent price level is without a doubt way below the social cost of carbon. It therefore has a limited impact on emissions. It even allowed electricity producers to substitute gas with coal, which emits 100 percent more carbon (not counting dirty microparticles) per kWh. An additional problem was that the ETS scheme covered only a fraction of the emissions of the region. Many specific emitters, e.g., the transport and building sectors, faced a zero-carbon price. During the third trading period (2013–2020), the EU-wide cap on emissions has been reduced by 1.74 percent each year, and a progressive shift toward auctioning of allowances in substitution for cost-free allocation has been implemented.

Over the past three decades, Europeans have sometimes believed that their (limited) commitment to reduce their emissions would motivate other countries to imitate their proactive behavior. That hope never materialized. Canada, for example, facing the prospect of the oil sands dividend, quickly realized that their failure to fulfill their commitment would expose them to the need to buy permits, and preferred to withdraw before having to pay for them. The U.S. Senate imposed a no-free-rider condition as a prerequisite for ratification, although the motivation for this otherwise reasonable stance may well have been a desire for inaction in view of a somewhat skeptical public opinion. Sadly enough, the Kyoto Protocol was a failure. Its architecture made it doomed to fail. Nonparticipating countries benefited from the efforts made by the participating ones, both in terms of reduced climate damages (free-rider problem) and in terms of improved competitiveness of their carbon-intensive industries (carbon leakage).

Other cap-and-trade mechanisms have been implemented since Kyoto. A mixture of collateral damage (we mentioned the emissions by coal plants of SO$_2$, a local pollutant, jointly with that of CO$_2$), the direct self-impact of CO$_2$ emissions for large countries like China (that has 20 percent of the world population and is exposed to serious climate change risk), and the desire to placate domestic opinion and avoid international pressure all lead to *some* carbon control. Outside the Kyoto Protocol, the United States, Canada, and China established some regional cap-and-trade mechanisms. In the United States, where the per capita GHG emissions are 2.5 times larger than in Europe and in China, two initiatives are worth mentioning. In the Regional Greenhouse Gas Initiative (RGGI), nine Northeast and Mid-Atlantic U.S. states created a common cap-and-trade market to limit the emissions of their electricity sector. Here also, the current carbon price is way too low at around $5/tCO$_2$ (up from the price floor level of $2/tCO$_2$ during the period 2010–2012). Over the period 2015–2020, the CO$_2$ cap will be reduced by 2.5 percent every year. The system will release extra carbon allowances if the carbon price on the market exceeds $6/tCO$_2$. A similar system exists in California to cover the electricity sector, large industrial plants, and more recently fuel distributors thereby covering more than 85 percent of the state's emissions of GHGs.[1] In 2014, China established seven regional cap-and-trade pilots, officially to prepare for the implementation of a national ETS scheme. The fragmented cap-and-trade systems described above cover almost 10 percent of worldwide emissions, and observed price levels are low. This is another illustration of the tragedy of commons. These regional or national ETS could be used in the future under any international commitment regime, either a universal carbon price or a cap-and-trade mechanism.

Some countries have implemented a carbon tax. The most aggressive country is Sweden, in which a carbon tax of approximately €100/tCO$_2$ has been implemented in 1991. France has fixed its own carbon tax at €14.5/tCO$_2$. Both of these taxes are used for various purposes, such as raising revenue or addressing congestion externalities and road safety. They also now can be used to comply with an international commitment to a cap-and-trade or a carbon price. Outside of Europe, some modest carbon taxes exist in Japan and Mexico, for example. Except for the Swedish case, these attempts put a carbon price that is far too low compared to the social cost of carbon (SCC).

3. Computing the Right Price Signal: The Social Cost of Carbon

As explained above, the solution to restore efficiency in the face of a global externality such as climate change is to expose all emitters of

CO_2 to the same carbon price, which should be equal to the marginal damage that is generated by these emissions. Over the past two decades, governments have commissioned estimates of the social cost of carbon. In the United States, the U.S. Interagency Working Group (U.S. IAWG 2013) proposed three different discount rates (2.5 percent, 3 percent, and 5 percent) to estimate the SCC. Using a 3 percent real discount rate, their estimation of the SCC is $32 in 2010, rising to $52 and $71, respectively, in 2030 and 2050. In France, the Commission Quinet (Quinet 2009) used a real discount rate of 4 percent, and recommended a price of carbon (/tCO_2) at €32 in 2010, rising to €100 in 2030 and between €150 and €350 in 2050.

Although the Fifth Assessment Report of the IPCC (IPCC 2014) does not contain much information about it, there is now a sizeable literature about the social cost of carbon. The problem is that there is no consensus among experts about how to value the marginal climate damages generated by CO_2. To illustrate the uncertainty affecting the SCC, we reproduce in figure 3.3 an analysis performed by Nordhaus (2011). He used his RICE integrated assessment model with uncertain parameters related to the discount rate and climate sensitiveness. Figure 3.3 reproduces the density function for the SCC of 2015, expressed in dollar per ton of carbon. Notice that 1 ton of carbon generates 3.7 tons of CO_2, so that the Nordhaus's mean estimate of

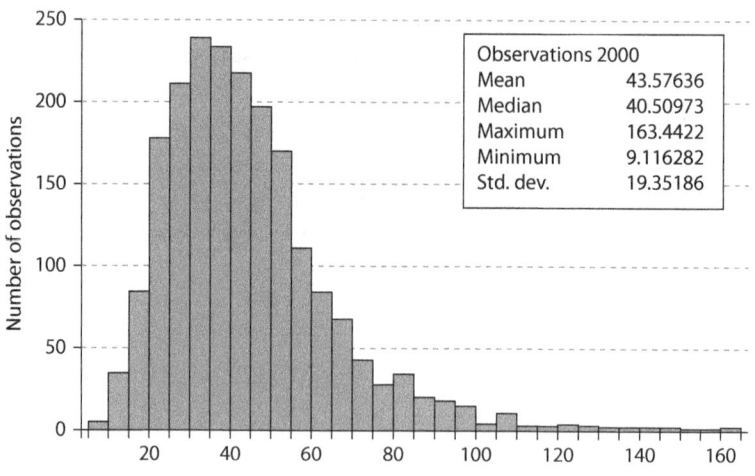

Figure 3.3 Social cost of carbon (in $/tC). *Source*: Nordhaus (2011).

the SCC at \$44/tC corresponds to \$12/tCO$_2$, which is considered relatively small compared to other estimates existing in the literature.

Three reasons can explain this disagreement. First, these damages are often nonmonetary, as they affect nontradable assets (natural capital, biodiversity, etc.). There is no agreed-upon methodology to value these assets. Second, these damages are uncertain. Climatologists are still debating about several aspects of their models: How much will the average temperature increase when one doubles the concentration of GHG in the atmosphere? Do we face the risk of a tipping point? How important are the stabilizing and destabilizing mechanisms of the earth's climate? Finally, most climate damage will materialize only in the distant future, which raises the question of our responsibilities toward future generations. These responsibilities are summarized in a key economic parameter that determines all arbitrages concerning immediate sacrifices and future benefits: the discount rate. Economists disagree about what rate should be used to discount the distant future.

In a growing economy, investing for the future—for example, by fighting climate change—raises intergenerational inequalities. Indeed, a global increase in investment in the economy forces the relatively poor current generation to reduce consumption, for the benefit of increasing the wealth of future generations, which are assumed to be wealthier than us. Keep in mind, for example, that with a 2 percent annual growth rate of consumption, people living two centuries from now will consume fifty times more goods and services than us, exactly as we consume fifty times more goods and services than our ancestors who lived two centuries ago. If we are inequality averse, investing for the future is socially desirable only if the return on these investments is large enough to compensate this adverse effect. This justifies using a positive discount rate. However, long-term growth is uncertain, and this justifies taking this intergenerational inequality argument with a pinch of salt. The modern asset pricing theory provides a useful benchmark to examine this valuation problem (Gollier 2012).

The argument translates into financial practice though the interest rate is positive. A simple arbitrage argument justifies using the interest rate to discount the safe benefits flow of any investment project. Indeed, there is a myriad of ways to improve the future, and it is crucial to prioritize them. Rather than fighting climate change, one could invest in the safe productive capital of the economy that will be used by future generations to increase their wealth. Fighting climate

TABLE 3.1
Real Return on 20-Year Government Bonds (annualized, in percent)

	2000–2014	1965–2014	1900–2014
China	3.0		
France	6.6	5.9	0.2
Germany	7.5	4.9	–1.4
Japan	33.9	4.4	–0.9
United Kingdom	3.6	3.2	1.6
United States	6.0	3.4	2.0
World	5.5	4.3	1.9

Source: Dimson, E., P. Marsh, and M. Staunton. 2015. *Credit Suisse Global Investment Returns Sourcebook 2015*. Zurich: Credit Suisse Research Institute.

change is desirable for them only if its "return" is larger than the return obtained by investing in the safe physical capital of the economy, which is measured by the interest rate. This explains why the interest rate is a good candidate for the discount rate, at least for safe investment projects. In table 3.1, I provide some information about the interest rates observed over the last 100 years.

From this table, assuming that future economic growth will be similar to the past, a relatively low real discount rate between 0 and 2 percent should be used to discount these damages to the present, if future climate damages were certain. This discount rate is in the order of magnitude of what has been used by Stern (2007) to estimate the SCC. He obtained an estimation of around $100 or more per ton of CO_2.

However, most standard integrated assessment models such as the DICE model are such that climate damages are positively linked to consumption growth (Dietz et al. 2018). For example, Nordhaus (2011) uses the outcome of Monte Carlo simulations of the RICE-2011 model with sixteen sources of uncertainty to conclude that "those states in which the global temperature increase is particularly high are also ones in which we are on average richer in the future" (21). Using technical terms from finance theory, this implies that the climate consumption-based CAPM beta is positive, and that the relevant climate discount rate is closer to the mean return of equity than to the risk-free rate (Gollier 2016). As seen in table 3.2 where I provide information about the mean return of equity during the last 100 years, a discount rate for

TABLE 3.2

Real Return on a Diversified Portfolio of Equity (annualized, in percent)

	2000–2014	1965–2014	1900–2014
China	3.0		
France	0.6	5.2	3.2
Germany	1.5	5.0	3.2
Japan	0.1	4.4	4.1
United Kingdom	1.0	6.2	5.3
United States	2.4	3.4	6.5
World	1.8	5.3	5.2

SCC around 3–5 percent seems to be more appropriate. This reduced discount rate typically reduces the SCC to around $40/tCO$_2$.

In sum, how much should be invested today to fight climate change depends upon how high the price of carbon should be. I have shown that determining the SCC strongly relies on our appreciation of our responsibilities toward future generations. This is an ethical question that transcends climate change, as it influences the way we should think about protecting biodiversity and nonrenewable resources, investing in infrastructure and education, or reducing public debt, for example. This ethical question translates into the choice of discount rate, which also depends upon the risk profile of climate damages.

4. Conclusion

Economists strongly recommend using carbon pricing to coordinate the planet in favor of an efficient fight for climate change. This recommendation has been endorsed by a wide range of companies around the world, such as Shell, ENGIE, or Microsoft, for example, that all use an internal price of carbon to drive their decisions. Their position is easy to understand: The worst economic climate for entrepreneurs is when there is unpredictability and unfair competition, and the smorgasbords of the command-and-control green policies have plenty of such ingredients. But, in December 2015, the COP21 in Paris buried carbon pricing amid a splendid general indifference. It suffices to say that Venezuela and Saudi Arabia considered this to be an unacceptable policy, thereby suggesting that these countries believe, along with prominent economists, that this only credible mechanism. The Paris Accord is extremely weak, with no formal commitment and no verification mechanism. Some countries have worked hard to make sure that this agreement contained no such coercive procedure, so that national pledges can be removed on short notice at no cost.

Investors and households do not believe that the Paris Accord will ever penalize high emitters of GHG. Otherwise, they would have massively disinvested from the least responsible companies around the planet to reinvest in the more virtuous ones. These should have been reflected in the asset prices. MSCI is an international company who has built various portfolios specializing in different sectors or regions. One of their indices tracks the value of a portfolio called ACWI LOW CARBON which, in each sector/country, biases its composition toward companies that emit less CO_2. In figure 3.4, I have drawn the evolution of the profit of an investor who gambled on the success of the COP21. On September 1, 2015, this investor would have taken a long-short position of $1 million, long on the ACWI LOW CARBON

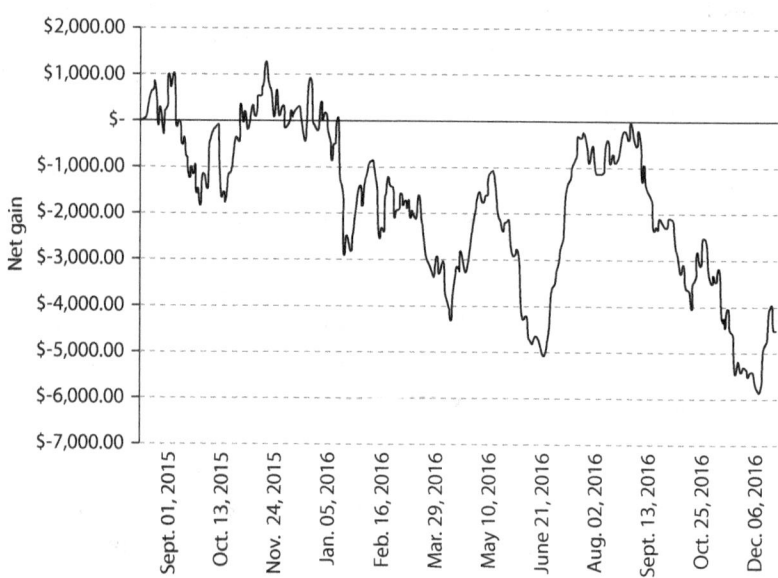

Figure 3.4 Profit from gambling on the success of the COP21.

Note: *Payoff of a $1 million long-short low carbon.*

portfolio, and short on the ACWI benchmark. If she would have liquidated the position during the Paris Conference, she could have netted a low profit of $1000. But as I write this, on January 12, 2017, this profit as been turned into a net loss of $4506!

This means that the possibility to put in place an efficient carbon-pricing policy remains a remote and somewhat unrealistic expectation. In any case, it is important to also think about what responsible actors can do. As explained in the previous section that a social cost of carbon, i.e., a carbon price around $40/tCO_2$ makes sense. This means that responsible investors should implement any action to reduce emission of CO_2 that costs them less than $40/tCO_2$. No more, no less.

Notes

The first two sections are inspired by Gollier and Tirole (2015).

1. Since early 2014, this market has been linked to a similar one established by the Province of Québec. The current price of permits in California is $12/tCO_2$, at the minimum legal price. This fragmented scheme illustrates the strange economics of climate change in the United States, where the minimum carbon price in California is larger than the maximum carbon price in Regional Greenhouse Gas Initiative (RGGI).

References

Dietz, S., C. Gollier, and L. Kessler. 2018. "The Climate Beta." *Journal of Environmental Economics and Management*, no. 87: 258–274.

Gollier, C. 2012. *Pricing the Planet's Future: The Economics of Discounting in an Uncertain World*. Princeton, New Jersey: Princeton University Press.

Gollier, C. 2016. "Evaluation of Long-Dated Assets : The Role of Parameter Uncertainty." *Journal of Monetary Economics*, no. 84: 66–83.

Gollier, C., and J. Tirole. 2015. "Negotiating Effective Institutions Against Climate Change." *Economics of Energy and Environmental Policy*, no. 4: 5–27.

Hardin, G. 1968. "The Tragedy of the Commons." *Science* 162 (3859): 1243–1248.

IPCC (Intergovernmental Panel on Climate Change). 2014. *Climate Change 2014: Mitigation of Climate Change*. Contribution of Working Group III to the Fifth Assessment Report of the Intergovernmental Panel on Climate Change (edited by Edenhofer, O., R. Pichs-Madruga, Y. Sokona, E. Farahani, S. Kadner, K. Seyboth, A. Adler, I. Baum, S. Brunner, P. Eickemeier, B. Kriemann, J. Savolainen, S. Schlömer, C. von Stechow, T. Zwickel, and J.C. Minx). Cambridge, UK/New York: Cambridge University Press.

Nordhaus, W. D. 2008. A Question of Balance: Weighing the Options on Global Warming Policies. New Haven, CT: Yale University Press.

Nordhaus, W. D. 2011. "Estimates of the Social Cost of Carbon: Background and Results of the RICE-2011 Model." Cowles Foundation Discussion Paper (October).

OECD (Organization for Economic Co-operation and Development). 2013. *Climate and Carbon: Aligning Prices and Policies*. OECD Environment Policy Paper 1. Paris: OECD Publishing.

Quinet, A. 2009. *La valeur tutélaire du carbone: Rapport de la commission présidée par Alain Quinet*. La Documentation Française, Rapports et Documents No. 16, Paris.

Stern, N. 2007. *The Economics of Climate Change: The Stern Review*. Cambridge, UK/New York: Cambridge University Press

U.S. Interagency Working Group (U.S. IAWG). 2013. "Technical Update of the Social Cost of Carbon for Regulatory Impact Analysis Under Executive Order 128666." Washington, DC: U.S. IAWG.

CHAPTER FOUR

How Should Countries Price Fossil Fuels?

IAN PARRY

THE 2015 PARIS Agreement on climate change, which came into force in November 2016 (following ratification by a sufficient number of countries), was a landmark achievement in international cooperation and global governance. There were 190 countries that submitted greenhouse gas (GHG) emission reduction pledges specified in their *nationally determined contributions* (NDCs), where a typical pledge will require cutting emissions by around 30 percent by 2030 relative to emissions in a baseline year (table 4.1, second column). These pledges are not enforceable—indeed, the United States intends to withdraw from the agreement—but countries are required to report every two years on their progress and to submit updated NDCs every five years, which are expected to be progressively more stringent.

At the same time, growing alarm about air pollution and other local environmental threats, recognition of the limitations of traditional environmental regulations, a preference for revenue-raising instruments given historically high fiscal pressures, and the window of opportunity created by lower energy prices, have all heightened interest in broader reform of fossil fuel prices.

TABLE 4.1
Emissions Pledges for Paris and the Impacts of Carbon Pricing, G20 Countries

| Country | Mitigation Pledge | 2014 | | | | 2030 | |
| | | Share of Global CO_2 | Tons $CO_2/\$GDP$ | Tons CO_2 per Capita | Required CO_2 Reduction, % | Effect of \$70/ ton CO_2 price | |
						CO_2 Reduction, %	Revenue, % GDP
Argentina	GHGs 15% below BAU in 2030	0.6	0.39	4.7	15	22	1.5
Australia	GHGs 26%–28% below 2005 by 2030	1.0	0.25	15.4	47	28	1.6
Brazil	GHGs 37% below 2005 by 2025	1.5	0.22	2.6	18	17	1.5
Canada	GHGs 30% below 2005 by 2030	1.5	0.30	15.1	45	22	2.2
China	CO_2/GDP 60%–65% below 2005 by 2030	28.5	0.98	7.5	24	30	2.7
France	GHGs 40% below 1990 by 2030	0.8	0.11	4.6	39	14	0.8
Germany	GHGs 40% below 1990 by 2030	2.0	0.19	8.9	36	30	1.2
India	GHG/GDP 33%–35% below 2005 by 2030	6.2	1.10	1.7	10	43	3.8

Country	Commitment						
Indonesia	GHGs 29% below BAU in 2030	1.3	0.52	1.8	15	22	2.5
Italy	GHGs 40% below 1990 by 2030	0.9	0.15	5.3	37	18	1.2
Japan	GHGs 25% below 2005 by 2030	3.4	0.25	9.5	27	20	1.5
Korea	GHGs 37% below BAU in 2030	1.6	0.42	11.6	37	29	2.1
Mexico	GHGs 25% below BAU in 2030	1.3	0.37	3.9	25	17	2.2
Russia	GHGs 25%–30% below 1990 by 2030	4.7	0.83	11.9	15	19	6.8
Saudi Arabia	GHGs 130 million tons below in 2025 and 2030	1.7	0.79	19.5	23	12	4.7
South Africa	GHGs 398–614 million tons in 2025 and 2030	1.4	1.40	9.0	32	36	7.7
Turkey	GHGs up to 21% below BAU by 2030	1.0	0.37	4.5	21	21	2.4
United Kingdom	GHGs 40% below 1990 by 2030	1.2	0.14	6.5	37	20	1.1
United States of America	GHGs 26%–28% below 2005 by 2025	14.5	0.30	16.5	33	28	1.5

Sources: UNFCCC (2017), WBG (2017), and preliminary International Monetary Fund (IMF) staff calculations.

To help guide reform, this chapter distills recent analyses of the efficient set of fossil fuel prices, focusing on G20 member countries, especially China (by far the largest emitter—table 4.1, third column). The starting point is that the efficient fuel price consists of the supply cost, the environmental costs—both global and domestic—and (if applicable) general sales taxes applied to products consumed at the household level. Most of the focus is on the environmental costs, as these are the least well understood.

Integrating Carbon Charges in Fossil Fuel Prices

Case for, and Design of, Carbon Pricing

It is widely accepted that carbon pricing should ideally be the principal instrument used to mitigate carbon emissions[1]—and for two reasons.

First, is that carbon pricing is easily the most effective policy instrument from an environmental perspective. As carbon prices are reflected in higher prices for fossil fuels, electricity, and so on, this provides across-the-board incentives for: switching among fuels with high (coal), intermediate (e.g., natural gas and coal combined with carbon capture and storage technologies), and zero (e.g., renewables, nuclear) carbon intensity; improvements in energy efficiency; and less use of energy-consuming products. In contrast, regulatory approaches, like requirements for renewables in power generation or improvements in the energy efficiency of vehicles or buildings, are by themselves far less effective as they focus on a much narrower range of emission reduction opportunities.

The second rationale for carbon pricing is that it raises potentially large amounts of (often badly needed) revenues.

In designing carbon pricing, there are three basics to get right.

The first is to comprehensively cover emissions. From an administrative perspective this is best done by imposing carbon charges upstream on the fossil fuel supply (from domestic and foreign sources) in proportion to the fuel's carbon content at the point of fuel extraction (e.g., mine mouth, oil well), processing (e.g., at coal washing plants or the refinery gate), or distribution.[2] Pricing fuels in this way as they enter the economy covers all potential emissions, while minimizing the number of collection points, and is a straightforward extension of existing fuel taxes (e.g., road fuel excises, royalties on extractives) that are well established in most countries and among the easiest of all taxes to administer.

The second basic design issue is productive use of the revenues from carbon pricing, most obviously, revenues can be used to lower the burden on the economy from broader taxes on labor and capital that distort formal work and investment incentives, to fund high-value (general or environmental) public spending or to reduce budget deficits. Forgoing these recycling benefits by, for example, using revenues for lump-sum dividends or projects with low benefit/cost ratios substantially increases the overall costs of carbon pricing for the economy and undermines the case for carbon pricing over other instruments.[3]

The third issue is to establish emissions price trajectories that are both predictable—as this is important for mobilizing investments in low-emission technologies—and in-line with environmental goals. Economists have generally recommended emissions prices reflect estimates of the *social cost of carbon* (SCC)—the present value of future global economic and environmental damage from additional climate change due to an additional ton of current CO_2 emissions[4]—though for practical purposes, emissions prices need to meet the Paris mitigation pledges, which seems to be the more relevant goal at present (see below).

Each of these basic design objectives suggests a general preference for carbon taxes over emissions trading systems (ETSs). If countries are moving ahead with ETSs (perhaps because mitigation policy is delegated to the environment rather than the finance ministry) in principle, they should be designed to look like taxes. This means combining them with taxes to cover fuels (e.g., for vehicles, buildings) that are usually excluded from ETSs (as the latter are usually confined to large industrial emissions sources); auctioning the allowances to raise revenue (and transferring proceeds to the finance ministry to compete with the full range of possible revenue uses); and using price ceilings and floors to promote price predictability. Clearly, political considerations (e.g., a reluctance to raise energy prices) can lead to compromises in instrument choice and policy design, so it is important for policy makers to have a quantitative sense of the advantages of well-designed carbon-pricing schemes over other policies to make informed choices.

Prices for Paris

Focusing on CO_2 emissions (which account for around three-quarters of GHGs and are the most practical emissions to price), the sixth and seventh columns of table 4.1 provide preliminary estimates of the emissions reductions below business-as-usual (BAU) levels in 2030 implied by NDCs in G20 counties, and the emissions reductions induced from phasing in a comprehensive $70 per ton CO_2 price by 2030. These estimates are based on projections of fossil fuel use in different sectors (power generation, road transport, industry, homes, and so on) and assumptions about the responsiveness of fuel use (from empirical studies and models of energy technology adoption) to carbon pricing.

The $70 price is much higher than needed to meet NDC targets in some countries (e.g., China, India, Indonesia), roughly sufficient in some others (e.g., Turkey, the United States), and falls well short of what is needed in still other cases (e.g., Australia, Canada, Italy, the United Kingdom). These differences in the needed emissions price reflect differences both in the stringency of commitments, which vary from 10 percent in India to over 40 percent in Australia and Canada as well as in the responsiveness of emissions to pricing, for example, the $70 per ton price reduces emissions by 30 percent or more in China, Germany, India, and South Africa that use a lot of coal but by less than 15 percent in France and Saudi Arabia.[5] Although estimates of the prices implied by the NDCs are inherently uncertain, it seems clear that they differ considerably across countries, which should increase interest in international price coordination mechanisms (see below).

A final point from table 4.1 (last column) is the large amounts of revenues that would be raised if a $70 carbon price were phased in. Revenues are around 1 to 3 percent of GDP in most cases by 2030, and even higher in countries like South Africa (which has the highest CO_2 intensity of GDP), underscoring the potential for quite radical restructuring of the tax system.

A Closer Look at China

China is especially important to consider given it accounts, by far, for the largest share of global CO_2 emissions (28.5 percent in 2014,

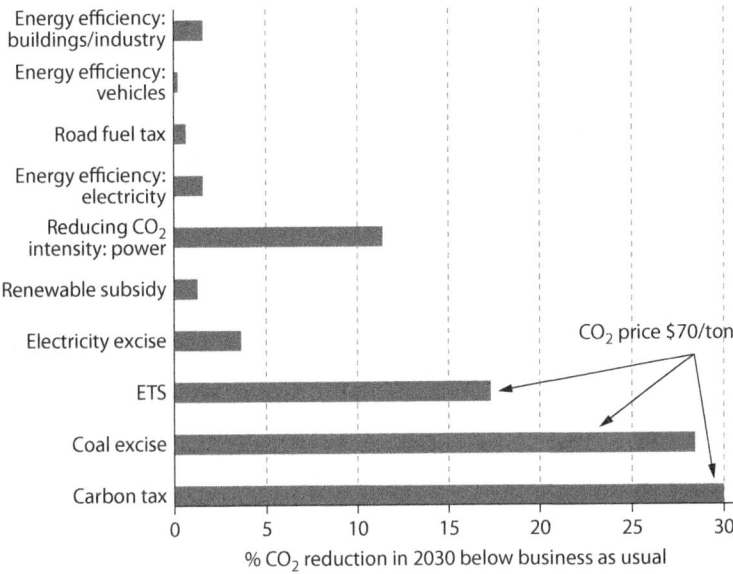

Figure 4.1 CO_2 effectiveness of alternative carbon mitigation instruments in China. *Source*: Parry et al. (2016).

compared with 14.5 percent for the United States, the second largest emitter—third column in table 4.1).

Figure 4.1 compares, for China, the effectiveness of a wide range of mitigation instruments at reducing CO_2 emissions in 2030. A carbon tax of $70 per ton reduces emissions by 30 percent below business-as-usual levels (i.e., with no mitigation policies). However, the huge bulk of the emissions reductions come from less coal use (rather than less petroleum and natural gas use), and therefore a tax on coal alone (imposing the same CO_2 price) achieves more than 90 percent of the emissions reductions under the carbon tax.

An ETS on the other hand is notably less effective, reducing CO_2 emissions by about 57 percent of the reductions under the (equivalently scaled) carbon tax, as the ETS does not cover small coal users in the industrial and residential sectors. Given that China plans to implement a nationwide ETS, this could be combined with an up-front tax on coal (a straightforward modification of the existing coal royalty collected at the mine mouth), perhaps with tax credits for large downstream coal users covered by the ETS (to avoid double charging).

Figure 4.1 also indicates, however, that the ETS is still a lot more effective than a range of other mitigation instruments, like energy efficiency policies, road fuel taxes, renewables subsidies, and incentives for reduced emissions intensity in power generation.

Also interesting for China is to consider the estimated effects of carbon policies on premature mortalities from exposure to local outdoor air pollution, as indicated in figure 4.2. In a business-as-usual case, mortalities are projected to rise from 1.1 million a year at present to about 1.3 million by 2030 (with rising energy use and urbanization outweighing greater deployment of emissions control technologies). The coal and carbon taxes would however reduce annual deaths to about 0.9 million in 2030, saving a substantial 3.7 million lives cumulated over the 2017–2030 period as they are progressively phased in. The ETS is significantly less effective, saving about 2 million lives.

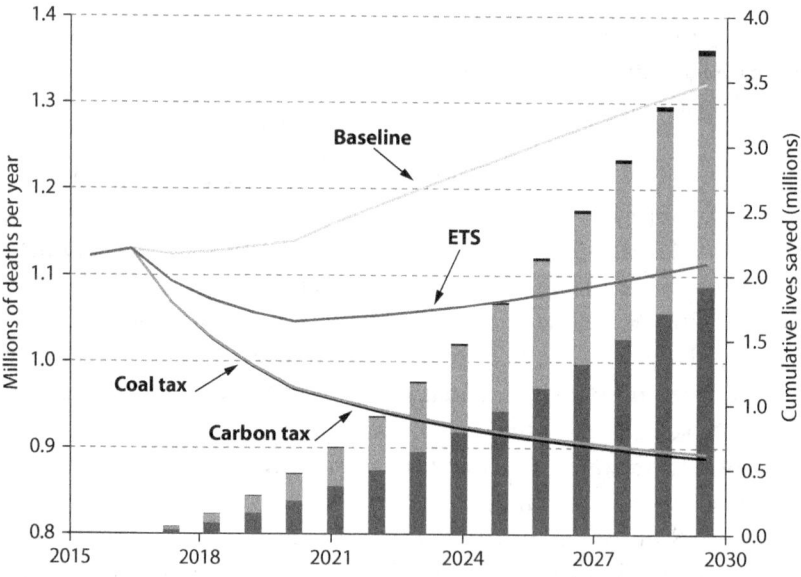

Figure 4.2 Local air pollution deaths in China under alternative carbon mitigation policies. *Source*: Parry et al. (2016).

Note: Results are for policies progressively phasing in a $70 CO_2 price by 2030. The curves (measured against the left axis) indicate the annual air pollution deaths under different policies while the boxes (right axis) indicate cumulative deaths saved under policies (the dark gray box for the ETS, the dark and light gray boxes for the coal tax, and the dark gray, light gray, and black boxes for the carbon tax).

Broader Energy Price Reform

Beyond carbon emissions, there are a range of other environmental externalities associated with fossil fuel use.[6] Most obviously, as just noted, is outdoor air pollution that is estimated to kill over 3 million people a year worldwide,[7] the main problem being fine particulates which are small enough to penetrate the lungs and bloodstream, thereby increasing the prevalence of various heart and lung diseases and strokes. Fine particulates can be emitted directly during fuel combustion or formed indirectly from chemical reactions in the atmosphere involving sulfur dioxide and nitrogen oxide emissions. The economically efficient policy here would be to directly tax emissions or (perhaps administratively more practical for some countries), impose up-front charges on fuel use combined with credits for downstream entities using emissions-control technologies. Either policy promotes the full range of behavioral responses for reducing emissions (use of mitigation technologies, shifting from high to low sulfur coal, shifting from coal and diesel to cleaner fuels, reducing demand for energy-using products).

More generally, the use of transportation fuels in road vehicles is associated with excessive traffic congestion, accidents, and wear and tear on the road network, as individual drivers do not, for example, account for their impact on slowing travel speeds for other road users or the injury risks they pose to pedestrians and other vehicle occupants. The more efficient policy here is taxes on miles driven (rather than fuel), for example, fees on busy roads rising and falling during the course of the rush hour promote the full range of congestion-reducing responses (including people leaving much earlier or later to avoid the rush hour peak, shifting to off-peak travel, switching to less congested roads, carpooling, shifting to transit and other travel modes).

However, until these ideal pricing policies are comprehensively implemented (likely a long time for most countries) it is entirely appropriate to reflect all environmental costs in fuel taxes—not doing so reduces the efficiency of the economy and leads to perverse policy implications (e.g., fuel taxes for European countries that are closer to U.S. levels).

Figure 4.3 provides some sense of the difference between current and efficient prices for coal, natural gas, gasoline, and road diesel for G20 member countries for 2013. These figures assume, for illustration,

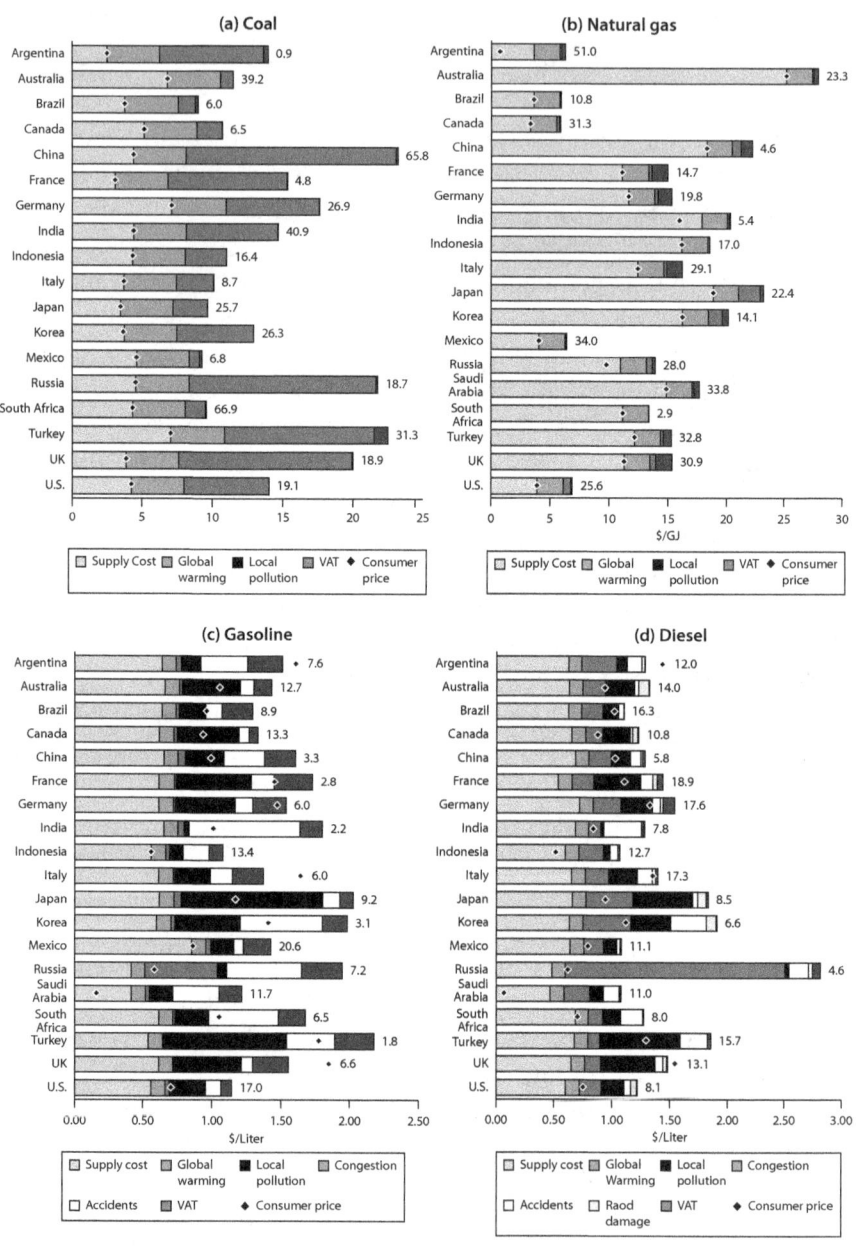

Figure 4.3 Current and efficient energy prices in G20 countries. *Source:* Coady et al. (2017), updated from Parry et al. (2014).

Note: Figures to the right of the bars are the share of fuels in primary energy.

a value of $40 per ton for future global warming damages associated with 2013 CO_2 emissions for all countries[8]—alternative values would imply proportionate changes in the carbon component of the efficient fuel price.

For coal, air pollution damages can be substantial, more than the carbon damages for ten countries, and more than double carbon damages in six cases. Not surprisingly however, air pollution damages differ dramatically across countries, being relatively high for example in densely populated China where a lot of people are exposed to the emissions, and relatively modest in sparsely populated Australia where the converse applies (and many plants are located coastally so some emissions disperse over the oceans without harming people). Nonetheless, the overall picture is one of pervasive and substantial undercharging for coal use, given that current prices essentially reflect only the supply costs, not environmental costs (though reform is less pressing where coal is a small share of primary energy, for example, in Argentina, Brazil, Canada, and France).

Taxes on natural gas are also rare and some countries (e.g., Argentina, India, Russia) modestly subsidized it in 2013, though overall the differentials between current and efficient prices for natural gas are far less striking than for coal. This is because natural gas is a much cleaner fuel—it does not produce direct fine particulate emissions or sulfur dioxide, and the per unit of energy nitrogen oxide and carbon emissions are a fraction of those for coal.

Gasoline also appears to be underpriced in most countries, in the sense that estimated prices need to cover supply costs and environmental costs, and general sales taxes exceed current prices in all but three G20 countries (Argentina, Italy, the United Kingdom) in figure 4.3, though congestion and accidents (which are better addressed through mileage-based taxes) tend to be much larger than carbon and air pollution damages. Undercharging for road diesel tends to be a little more pronounced than for gasoline, as the fuel is often taxed at a lower rate than gasoline, and some environmental costs (e.g., local air pollution) can be larger.

There is much at stake in getting energy prices right. At a global level, Coady et al. (2015) estimate that had fossil fuel prices fully reflected supply and environmental costs (and general consumer taxes) in 2013, global CO_2 emissions would have been 21 percent lower than they would have been under actual fuel prices, fossil fuel air pollution deaths 55 percent lower, government revenues 4 percent higher, and there would have been a net economic gain (environmental benefits less economic costs) of 2 percent of gross domestic product (GDP).

Moving Policy Forward

To move energy pricing reform forward at a domestic level, policy makers need to assist the transitions away from activities that become uneconomic with energy efficiently priced, for example, through worker retraining and relocation and gradual, widely anticipated price reforms. Energy-intensive, trade-exposed firms are often granted favorable provisions under carbon pricing,[9] though ideally these would be scaled back with international proliferation of pricing schemes. Low-income households also need to be protected (e.g., through broader and deeper social safety nets), a rough rule of thumb being that well-targeted programs require around 10 percent or less of the revenue from fuel price reform (because 90 percent or more of the burden of higher prices is typically borne by the nonpoor).[10]

Policy makers also need to communicate the environmental (e.g., headway on Paris commitments) and fiscal (e.g., what other taxes will be cut or what spending will be prioritized) benefits of fuel price reform to legislators and the general public. A point to emphasize here is that a large portion of the reform benefits may be national (as prices are pushed up to reflect domestic environmental costs and revenue needs), so up to a point, countries can move ahead unilaterally and make themselves better off—they do not need to wait for other countries to act.

These domestic actions might be subsequently strengthened and coordinated at the international level through carbon-price floor arrangements among large emitters. These arrangements (that could complement the Paris process) provide more price certainty and some reassurance against concerns that countries will lose international competitiveness while allowing countries the flexibility to set carbon prices higher than the floor price, which they may wish to do if they have high fiscal needs from carbon taxes, relatively large domestic environmental benefits, or because high prices are needed to meet Paris mitigation pledges. Precedents for these arrangements include tax floor treaties in the European Union for value-added taxes, and excises on tobacco, alcohol, and energy products (though a lesson here is that it is easier to make progress when the initial group of countries is small). Broader changes in energy taxes and subsidies that can undermine (or enhance) the effectiveness of a direct carbon price would need to be taken into account in these arrangements, as well as special reliefs from carbon pricing for sensitive sectors, but the analytical challenges for tracking countries overall *effective* carbon prices should be manageable.[11]

Although much is happening on the ground, with carbon-pricing schemes proliferating at national and subnational levels, and fuel price liberalizations in numerous energy producers (table 4.2) that seemed

TABLE 4.2
Recent Energy Price Reforms: Selected Countries

Country	Reform
Angola	Liberalize domestic fuel prices by 2020
Egypt	Fuel and gas prices increased 40%–78%, electricity prices 20%–50% in 2014
Ghana	Petroleum prices liberalized in 2015
Haiti	Gasoline, diesel, kerosene prices increased 6%–8% in 2014, 9%–11% in 2015
India	Gasoline prices liberalized in 2010 and diesel prices in 2014
Indonesia	Abolished gasoline subsidies and capped diesel subsidies in 2014
Jordan	Automatic pricing mechanism in 2012, fuel subsidies zero in 2014
Kuwait	Raised diesel and kerosene prices 210% in 2015 (partially reversed)
Madagascar	Eliminated fuel subsidies and implemented automatic pricing in 2016
Malaysia	Prices for gasoline and diesel set monthly to reflect international prices
Mexico	Domestic fuel prices liberalized in 2018
Morocco	Gasoline, diesel, industrial fuel oil, and liquefied petroleum gas subsidies eliminated
Saudi Arabia	Gasoline price increased 50% in 2015, planned increases for diesel, gas, electricity
Sudan	Plan to eliminate fuel subsidies by 2019 (but fuel price riots in 2013)
United Arab Emirates	Fuel price mechanism in 2015 and gasoline/diesel prices increased 25%–30%
Yemen	Gasoline, diesel, kerosene prices increased 20%, 50%, 100%, respectively in 2014

Source: IMF staff collected from country authorities.

a remote prospect just a few years ago, fuel prices are still far short of where they need to be—for example, the globally averaged (formal) price on CO_2 emissions was barely $1 per ton in 2016.[12] But the stakes from continued reform have never been higher—without a robust price signal to help redirect long-term investment toward low-emission technologies, fifteen years from now the world will likely be locked into a higher level of global warming than deemed safe by the architects of the Paris Agreement.

Notes

1. See for example, Farid et al. (2016), Krupnick et al. (2010), and Carbon Pricing Leadership Coalition (www.carbonpricingleadership.org/carbon-pricing -panel). Other policies are needed, particularly to accelerate the development of cleaner technologies, but the main focus here is on fossil fuel pricing.
2. See for example, Calder (2015).
3. See for example, Parry (2015).
4. See U.S. IAWG (2013).
5. See Aldy et al. (2016) for further discussion of needed emissions prices.
6. Parry et al. (2014).
7. See Lelieveld (2015) for further discussion.
8. Based on U.S. IAWG (2013).
9. See for example, Environment and Climate Change Canada (2017).
10. Coady, Flamini, and Sears (2015).
11. By recognizing internationally traded mitigation units, Article 6.2 of the Paris Agreement provides a potential vehicle for encouraging participation in price floor agreements. For example, countries that meet their emissions commitments with prices well below the floor price may still participate as they can gain from exceeding their commitments and selling the excess mitigation units at the floor price to other participants needing much higher prices to meet their NDCs.
12. World Bank Group (WBG 2016).

References

Aldy, Joseph, William Pizer, Massimo Tavoni, Lara Aleluia Reis, Keigo Akimoto, Geoffrey Blanford, Carlo Carraro, et al. 2016. "Economic Tools to Promote Transparency and Comparability in the Paris Agreement." *Nature Climate Change*, no. 6, 1000–1004.

Aldy, Joseph E., and William A. Pizer. 2015. "Comparing Countries' Climate Mitigation Efforts." In *Implementing a U.S. Carbon Tax: Challenges and Debates*, edited by I. Parry, A. Morris, and R. Williams, 233–252. New York: Routledge.

Calder, Jack. 2015. "Administration of a U.S. Carbon Tax." In *Implementing a U.S. Carbon Tax: Challenges and Debates*, edited by I. Parry, A. Morris, and R. Williams. New York: Routledge.

Coady, David, Valentina Flamini, and Louis Sears. 2015. "The Unequal Benefits of Fuel Subsidies Revisited: Evidence for Developing Countries." Working Paper 15/250. Washington, DC: International Monetary Fund.

Coady, David, Ian Parry, and Baoping Shang. 2018. "Energy Price Reform: A Guide for Policymakers." *Review of Environmental Economics and Policy*, forthcoming.

Coady, David, Ian Parry, Louis Sears, and Baoping Shang, 2015. "How Large Are Global Energy Subsidies?" Working Paper 15/105. Washington, DC: International Monetary Fund.

Environment and Climate Change Canada. 2017. *Technical Paper: Federal Carbon Pricing Backstop*. Canada: Gatineau QC.

Farid, Mai, Michael Keen, Michael Papaioannou, Ian Parry, Catherine Pattillo, and Anna Ter-Martirosyan. 2016. "After Paris: Fiscal, Macroeconomic, and Financial Implications of Climate Change." Staff Discussion Note 16/01. Washington, DC: International Monetary Fund.

IEA (International Energy Agency). 2016. *World Energy Balances*. Paris: IEA.

Krupnick, Alan J., Ian W.H. Parry, Margaret Walls, Tony Knowles, and Kristin Hayes. 2010. "Toward a New National Energy Policy: Assessing the Options." Washington DC: Resources for the Future and National Energy Policy Institute.

Lelieveld, J., J. S. Evans, M. Fnais, D. Giannadaki, and A. Pozzer. 2015. "The Contribution of Outdoor Air Pollution Sources to Premature Mortality on a Global Scale." *Nature* 525 (September 17): 367–371. www.nature.com /nature/journal/v525/n7569/abs/nature15371.html.

Parry, Ian, Baoping Shang, Philippe Wingender, Nate Vernon, and Tarun Narasimhan. 2016. "Climate Mitigation in China: Which Policies Are Most Effective?" Working Paper 16–148. Washington, DC: International Monetary Fund.

Parry, Ian W.H., Dirk Heine, Shanjun Li, and Eliza Lis. 2014. *Getting Energy Prices Right: From Principle to Practice*. Washington, DC: International Monetary Fund.

Parry, Ian. 2015. "Choosing Among Mitigation Instruments: How Strong Is the Case for a U.S. Carbon Tax?" In *Implementing a U.S. Carbon Tax: Challenges and Debates*, edited by I. Parry, A. Morris, and R. Williams. New York: Routledge.

UNFCCC. 2017. *INDCs as Communicated by Parties.* UN Framework Convention on Climate Change. http://www4.unfccc.int/submissions /indc/Submission%20Pages/submissions.aspx.

U.S. IAWG (United States Interagency Working Group). 2013. "Technical Update of the Social Cost of Carbon for Regulatory Impact Analysis Under Executive Order 12866." Washington, DC: U.S. IAWG.

WBG (World Bank Group). 2016. *State and Trends of Carbon Pricing 2016.* Washington, DC: World Bank Group.

WBG (World Bank Group). 2017. *World Bank Indicators.* Washington, DC: World Bank Group. https://data.worldbank.org/indicator.

CHAPTER FIVE

Should Carbon Pricing Be Different Across Countries?

KATHELINE SCHUBERT

Introduction

Energy taxes are very different across countries, as illustrated in figure 5.1 where the average effective tax rate on energy ranges from almost zero in Indonesia or Russia to more than €6/GJ in Denmark, and can even be very different across similar countries. Moreover, in each country and on average across countries, taxes are very different across products and uses, as illustrated in table 5.1. What is actually taxed is oil for transport, whereas coal is rarely taxed. Even oil taxes are very different across countries (see figure 5.2). These patterns are persistent over time, as noted by Newbery (2005) and OECD (2015). To be convinced that this apparent disorder is not totally irrational, let us examine the reasons why we tax or should tax energy.

First, energy taxation is used to finance public spending. We will not get into the famous debate on direct versus indirect taxation (see Atkinson and Stiglitz 1976), nor the debate over the uniform versus nonuniform taxation of goods. Instead, let us underline that

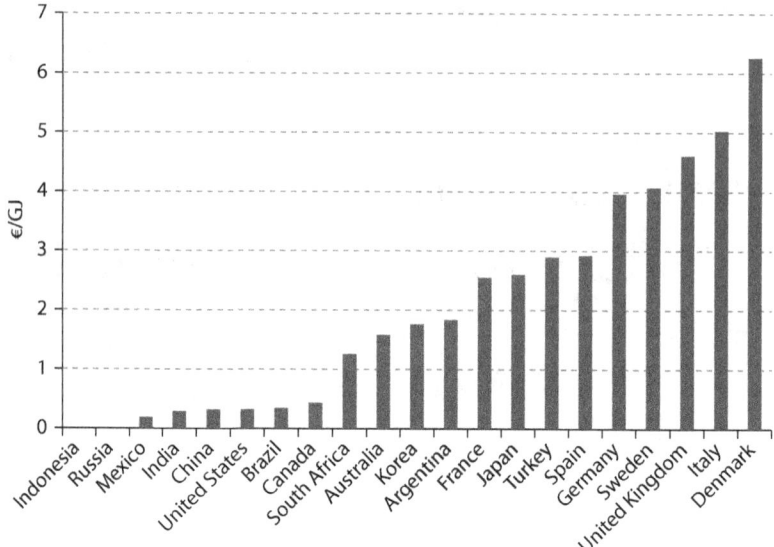

Figure 5.1 Economy-wide average effective tax rates on energy in selected countries (€/GJ). *Source: Taxing Energy Use 2015* (OECD 2015).

energy taxes, under the form of excise taxes, were originally purely used as revenue-raising instruments, without environmental purposes. Second, fossil energy is taxed to internalize negative externalities, so that they are reflected in energy prices and energy users change their consumption behavior accordingly. This is the famous Pigouvian part of energy taxes (Pigou 1920). The negative externalities considered can be created directly by energy combustion, as is the case for local air pollution or the climate change externality. Alternatively, energy taxation can be used as a second-best instrument for charging for transport infrastructure or correcting congestion externalities.

The different motives for taxing energy are of course difficult to disentangle. But a close look at the data teaches us first that countries tend to tax imported fossil energy to reap a part of the producers' scarcity rent (Bergstrom 1982), and second that Pigouvian taxation is (almost) everywhere underdeveloped. The direct measure of external effects is difficult and approximate, but an indirect proof of the second point is the undertaxation of coal, which is the more polluting of fossil energies, locally and also globally. When Pigouvian taxation exists,

TABLE 5.1
Weighted Average Effective Tax Rates on Energy by Fuel Type and Use (€/GJ), 41 Countries

	Oil Products	Coal and Peat	Natural Gas	Biofuels and Waste	Renewables and Nuclear	All Fuels
Percent of base	27 percent	34 percent	20 percent	9 percent	11 percent	100 percent
Transport use	5.20	0.00	0.12	3.74	0.00	4.96
Heating and process use	0.82	0.05	0.21	0.00	0.00	0.26
Electricity production	0.50	0.13	0.43	0.65	0.38	0.27
Total use	3.52	0.10	0.28	0.30	0.38	1.11

Source: Energy Use 2015: OECD and Selected Partner Economics, OECD.

Note: 41 countries: 34 OECD countries, plus Argentina, Brazil, China, India, Indonesia, Russia, and South Africa.

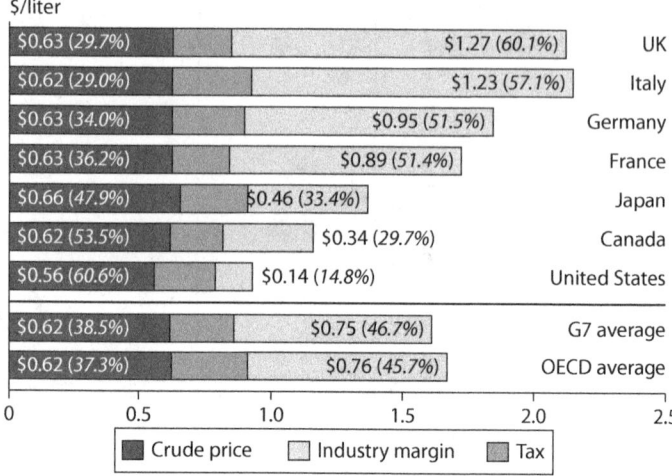

Figure 5.2 Taxes on oil: Who gets what from a liter of oil in 2014? *Source:* http://asb.opec.org/index.php/taxes-on-oil.

it is mostly designed to internalize local externalities. This last point is easy to prove because when governments put in place a carbon tax they publicize it widely. However, taxing carbon to curb emissions and fight climate change is an urgent and pressing issue.

Now, how should carbon taxation be designed?

The virtues of a uniform carbon price, taking the form of a tax or a price of tradable emission permits, reflecting the social cost of carbon, are well known: it allows for the equalization of marginal abatement costs and therefore minimizes the worldwide total cost of abating emissions. But we have to admit that a uniform carbon tax has not been implemented yet, and that we are very far from it: carbon taxation is not even mentioned in the COP21 agreement.

The standard theory argument is valid in a first-best setting, and neglects the existence of non-Pigouvian motives of taxation and of preexisting taxes, thereby ignoring tax interactions. However, in the real world, as we have seen above, there are big differences across countries, fuels, and uses in initial (pre-carbon) energy taxation. Should we superimpose the same carbon tax everywhere, knowing that taxes on energy intended to curb local externalities or finance public spending do a part of the job of the carbon tax, even if they were not intended to?

There are also big differences across countries in purchasing power, and different historical responsibilities. Can equity and climate justice concerns lead to the acceptance of different carbon taxes? We propose in what follows an answer to these questions.

1. The Carbon Tax at the First, Second, and Third Bests

The model we use is based on d'Autume, Schubert, and Withagen (2016). It is a very simple static model. The world is composed of n countries that interact through global emissions and possibly inter-country transfers. There is one representative consumer per country, which means that intracountry redistribution issues are ignored. There are three consumption goods: C, a private and nonpolluting generic good; X, a private and polluting good; and G, a public and nonpolluting good. Emissions of country i are directly equal to its consumption of the polluting good: $Z_i = X_i$. Similarly, world emissions are the sum of the consumptions of the polluting good: $Z_w = \sum_{i=1}^{n} Z_i$. The utility function of the representative consumer in country i reads $U^i(C_i, X_i, G_i, Z_i, Z_w)$, where the function U^i has all the usual properties. There are no strategic behaviors.

1.1 First Best

Imagine first that there exists a world social planner who maximizes a weighted sum of the utilities of the different counties of the world, subject to the resource constraint specifying that the sum of the exogenous endowments of each country is at least equal to the sum of the consumptions of the three goods. The weight β_i the planner gives to country i, or more precisely to the consumers of country i since countries play no role at this stage, is determined according to equity principles, or to take care of historical responsibilities in climate change (climate justice), or according to any other rule the planner decides to follow.

Classically, optimality conditions state that the world planner's choice of the consumptions of the three goods is such that weighted marginal utilities from clean consumption are equalized across countries:

$$\beta_1 U_C^1 = \ldots = \beta_i U_C^i = \ldots = \beta_n U_C^n$$

and that the marginal rate of substitution between the three goods is equal to their relative production costs (here, = 1):

$$\frac{U_G^i}{U_C^i} = 1, \quad i = 1, \ldots, n$$

$$\frac{U_X^i}{U_C^i} + \frac{U_Z^i}{U_C^i} + \sum_{j=1}^{n} \frac{U_{Z_w}^j}{U_C^j} = 1, \quad i = 1, \ldots, n$$

where U_C^i is the partial derivative of the utility function U^i with respect to C_i, and the other expressions have a similar meaning.

This first best may be decentralized. Local governments, at the country level, impose a local Pigouvian tax aimed at correcting the local externality (at rate ϕ_i in country i), while a global carbon tax at rate τ is implemented at the world level by an International Environmental Agency (IEA). The IEA redistributes the sum of carbon tax receipts, $\Sigma_i \tau X_i$, to local governments, each government being given a transfer T_i that does not necessarily correspond to the carbon tax paid by its consumers. Each of the local governments uses this transfer, together with local tax receipts $\phi_i X_i$, to finance local public goods provision G_i and lump-sum transfers to its consumers T_{ci}. Nothing prevents these transfers to consumers from being negative.

D'Autume, Schubert, and Withagen (2016) show that the tax rates ϕ_i and τ paid by consumers in country i read:

$$\phi_i = -\frac{U_Z^i}{U_C^i}, \quad \tau = -\sum_{j=1}^{n} \frac{U_{Z_w}^j}{U_C^j}$$

The local Pigouvian tax ϕ_i is equal to the marginal local damage due to the consumption of the polluting good, whereas the global part τ is equal to the sum of all worldwide marginal damages resulting from global warming.

1.2 Second Best

Suppose now that, in the decentralized setting, local governments are unable to finance the public goods provision through lump-sum taxation of consumers, as in the Ramsey (1927) optimal taxation approach. The new constraint reads $T_{ci} \geq 0$, $i = 1,...,n$. As was first shown by Sandmo (1975), the presence of externalities modifies the optimal tax scheme. If Pigouvian tax receipts are sufficient to finance the public goods provision and if the desired redistribution is limited, the first-best allocation is attainable. If not, additional distortionary taxation is required and we switch to a second-best situation. A new variable appears, reflecting the cost of being unable to levy a lump-sum tax: the cost of public funds μ_i.

In the standard Ramsey model, the main determinant of the cost of public funds is the price elasticity of demand for the taxed good. A high elasticity implies that high tax rates are necessary to obtain a given amount of funds. Then activity is reduced, and the cost of public funds is high. With externalities, the cost of public funds is reduced because emissions taxes decrease pollution and provide the government with receipts, alleviating the cost of obtaining funds. Here, the cost of public funds can even become zero if emissions tax receipts are sufficient to finance public spending.

D'Autume, Schubert, and Withagen (2016) show that in this second-best configuration the carbon tax is still uniform. The overall tax on the polluting good is decomposed in two country-specific local taxes and the global carbon tax:

$$\phi_i = -\frac{1+\mu_i H_Z^i}{1+\mu_i+\mu_i H_C^i} \frac{U_Z^i}{U_C^i}, \qquad \psi_i = \mu_i \frac{H_C^i - H_X^i}{1+\mu_i+\mu_i H_C^i} \frac{U_X^i}{U_C^i}$$

$$\tau = -\sum_{j=1}^{n} \frac{1+\mu_j H_{X_w}^j}{1+\mu_j+\mu_j H_C^j} \frac{U_{Z_w}^j}{U_C^j}$$

where the H terms are complex interaction terms depending on second-order derivatives of the utility function. Tax ϕ_i is again a local Pigouvian tax. Tax ψ_i is a Ramsey tax designed to finance the local public goods provision. Tax τ is the world carbon tax. The cost of public funds plays an important role in the formulas, besides the interaction terms. It is not possible to infer from the formulas whether ϕ_i and τ are higher or smaller than at the first best, and we will resort to numerical simulations to get insights on that matter.

1.3 Third Best

Suppose finally that for political economy reasons, local governments are not ready to accept a smaller transfer than the amount of carbon taxes their citizens are paying to the IEA. We impose a new constraint: $T_i \geq \tau_i X_i$, where τ_i is the carbon tax in country i. We leave open the possibility of different carbon taxes across countries. This configuration is called third best.

Generalizing Chichilnisky and Heal's (1994) result, D'Autume, Schubert, and Withagen (2016) show that indeed the carbon tax is no longer uniform, and reads:

$$\tau_i = -\frac{\sum_{j=1}^{n} \beta_j \left(1 + \mu_j H_{X_w}^j\right) U_{Z_w}^j}{\beta_i \left(1 + \mu_i + \mu_i H_C^i\right) U_C^i}$$

Moreover, the formula of the Pigouvian and Ramsey tax rates ϕ_i and ψ_i are the same as at the second best, which of course does not mean that the level of these taxes are the same, as they are computed at different equilibria.

To get insights on these different carbon taxes, suppose that the social weights β_i reflect the IEA's aversion to inequality. Then a poor country will be given a high β. As this country will have a high marginal utility of consumption and probably also a high cost of public funds, its coefficient $1 + \mu + \mu H_C$ will be high. Consequently, its carbon tax will be low.

2. Illustration in the Two-Country Case

We now illustrate the previous results in the simple case of two countries. We suppose that utility functions are separable, linear in the consumption of the generic good, isoelastic in the consumption of the polluting and the public goods, and that local and global damages are also isoelastic. They read:

$$U^i(C_i, X_i, G_i, Z_i, Z_w) = C_i + \alpha_i^x \frac{X_i^{1-\frac{1}{\varepsilon_i^x}}}{1-\frac{1}{\varepsilon_i^x}} + \alpha_i^g \frac{G_i^{1-\frac{1}{\varepsilon_i^g}}}{1-\frac{1}{\varepsilon_i^g}} - \alpha_i^z \frac{Z_i^{1+\frac{1}{\varepsilon_i^z}}}{1+\frac{1}{\varepsilon_i^z}}$$

$$- \alpha_i^{z_w} \frac{Z_w^{1+\frac{1}{\varepsilon_i^{z_w}}}}{1+\frac{1}{\varepsilon_i^{z_w}}}, \quad i = 1, 2$$

where all coefficients are positive.

We begin by studying the case of two identical countries.

First, we suppose that the IEA gives the same weight to the two countries in the social welfare function ($\beta_1 = \beta_2 = 1$), and compare the economic outcomes of the situation without a carbon tax (the business-as-usual situation, BAU) and the situation with a carbon tax. The introduction of the carbon tax does not modify the theoretical formula characterizing local taxes (see above). However, it does modify the level of these taxes. We assess to what extent it does so, at the first best and at the second best.

Second, assuming now that the carbon tax exists, we compare the first-best and the second-best solutions, for different weights granted by the IEA to country 2 compared to country 1 ($\beta_2 \neq \beta_1 = 1$).

Third, assuming again that the carbon tax exists, we examine to what extent the third best is more constrained, in a sense that will be clearly defined, than the second best.

We then depart from the assumption of symmetric countries and suppose that consumers of country 1 value public goods more than consumers of country 2, everything else being equal. We consider the case where the IEA gives the same weight to the two countries in the social welfare function ($\beta_1 = \beta_2 = 1$), and compare the first best, the second best, and the third best.

2.1 Identical Countries

Suppose first that the two countries are identical. The purely illustrative calibration we choose is such that at the first best, tax receipts are not sufficient to finance public spending at the world level. Hence, at the first best, the sum of lump-sum transfers to consumers is negative. Then one or both countries will be constrained at the second best,

TABLE 5.2

Efficient Equilibrium and Second Best for Equal Weights and Identical Countries

			Efficient Equilibrium					
	X_i	G_i	T_{ci}	ϕ_i	τ	$\phi_i + \tau$	T_i	μ_i
BAU	36.4	20	−13	0.191	0	0.191	0	0
Carbon tax	28.6	20	−7	0.085	0.372	0.457	10.6	0

			Second Best					
	X_i	G_i	ψ_i	ϕ_i	τ	$\psi_i + \phi_i + \tau$	T_i	μ_i
BAU	29.1	12.7	0.376	0.062	0	0.438	0	0.46
Carbon tax	27	14.2	0.314	0.053	0.160	0.527	4.3	0.33

have a positive cost of public funds, and will have to levy distortionary taxes to finance public spending.

Table 5.2 summarizes the outcomes of the BAU situation (no carbon tax) and of the situation with carbon tax and intercountry transfers, when the International Environmental Agency gives equal weights to the two countries in the social welfare function ($\beta_1 = \beta_2 = 1$). The BAU is a situation where all the decisions are taken by the local governments, who neglect their impact on global pollution. They impose a welfare maximizing local pollution tax. At the efficient BAU the local governments can impose lump-sum taxes to their citizens, whereas at the second-best BAU lump-sum taxes are not feasible.

Compare first the BAU and the simulation with the carbon tax, at the first best or at the second best (the following remarks apply in both situations). The interesting point is that local taxes on the polluting good (ϕ_i) must be reduced when the world carbon tax on this same good is introduced. The explanation is straightforward: the carbon tax, levied on the same good as the local tax, makes a part of the job of reducing local pollution, even if it is not intended for that. Nevertheless, total taxes ($\phi_i + \tau$) on the polluting good are higher with than without the carbon tax, as now more externalities are internalized (the local plus the global one). Hence, consumption of the polluting good (X_i) decreases.

It may also be noted that when the carbon tax is introduced, pollution tax receipts ($\phi_i + \tau$)X_i are higher, which reduces the need to resort to lump-sum taxes (T_{ci}) at the first best or additional distortive taxes (ψ_i) at the second best, in order to finance public spending. At the second best, the marginal cost of public funds is lower with the carbon tax than without, and this results in higher public spending: the carbon tax, aimed at curbing the global warming externality, also helps to finance public spending.

Compare now the first-best and the second-best outcomes in the case with carbon tax. The marginal utility of public spending is equal to its total cost, namely 1 at the first best, and 1 plus the marginal value of public funds at the second best. Therefore, public spending is smaller at the second best than at the first best. Moreover, in a first-best framework, the public goods provision is financed through lump-sum taxes and the only role of commodity taxation is to limit emissions, while in a second-best framework, commodity taxation has to achieve the two objectives. A new component ψ_i is added to ϕ_i and τ. Therefore total taxes are higher, inducing a lower level of emissions, and consequently lower local and global damages. This is the reason why even if total taxes are higher, the Pigouvian taxes ϕ_i and τ are smaller at the second best than at the first best.

We now turn to the situation where the IEA wants to favor one or the other country, and compare the first-best and the second-best solutions, for different values of β_2, β_1 being set to 1. Of course, the reason why the IEA could want to favor one country at the expense of the other in the symmetric case is not obvious. We study this situation to disentangle the *pure* effect of different βs from the effects resulting from differences in endowments and/or preferences between the two countries.

When β_2 is very small compared to β_1, the IEA favors country 1 a lot, so that country 1 is not constrained and its marginal value of public funds μ_1 equals 0, whereas μ_2 is very high. For intermediate values of β_2, both countries are constrained and have a positive marginal cost of public funds. For higher values of β_2, $\mu_2 = 0$ and μ_1 is high. Remember that the calibration chosen is such that at the first-best, the sum of transfers to consumers is negative, so that there exists no second-best equilibrium with $\mu_i = 0$, $i = 1, 2$.

If the IEA wants to favor country 2 and to increase U^2, which means $\beta_2 > 1$, it increases its transfers T_2. This allows government 2 to tax less, to benefit from a lower cost of public funds, and possibly to make a lump-sum transfer to its citizens. The lower cost of public funds leads to higher public spending in country 2. The smaller emissions tax leads to a larger production of the polluting good X_2. Symmetrically, we have a decrease of pollution in country 1, which is relatively less favored by the IEA. We have seen above that when $\beta_2 = \beta_1$, world emissions are lower at the second best than at the first best. It is still the case when $\beta_2 > 1$, which means that the decrease of pollution in country 1 more than compensates the increase in country 2. Moreover, the further β_2 is from 1, the smaller total emissions are compared to their first-best level. The intuition behind this is that as the two countries are identical, $\beta_2 \neq 1$ (> 1 here) means more distortions, a higher average marginal cost of public funds of the two countries, hence higher average total taxes and lower world emissions. As the global damage is reduced, so is the carbon tax.

Finally, figure 5.3 shows the set of optima in the first-best, second-best, and third-best cases. Remember that the two countries are identical. Hence, for $\beta_2 = \beta_1 = 1$ the second-best and the first-best solutions are the same, and correspond to the case where the IEA redistributes to each country exactly the amount of carbon tax it has paid. The striking result is that the third best is much more constrained than the second best, in the sense that the range of utilities attainable is much smaller, and that within this range the IEA has to accept a large decrease of the utility of one country if it wants to obtain a small increase of the utility of the other one. Thus, the answer to the question posed by Sandmo (2004): "Should one design compensatory transfers, or should the design of the environmental taxes themselves have built-in

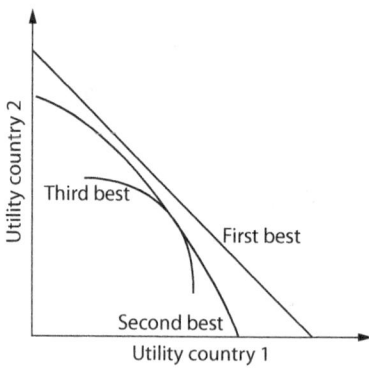

Figure 5.3 Utility frontiers: first, second, and third best.

distributional elements?" seems to be clearly that designing compensatory transfers (second-best case) is preferable than introducing into the tax itself distributional elements (third-best case). Whether this solution is politically feasible remains an open question.

2.2 Country 1 Values Public Spending More Than Country 2

In this simulation, countries are identical except for the weights of the utility derived from public spending. The assumption here is that country 1 values public spending more than country 2: $\alpha_1^g > \alpha_2^g$. Table 5.3 summarizes the results of the first-best, second-best, and third-best cases, when the IEA gives the same weight to both countries in the social welfare function. Here, the first best and the second best differ because at the first best the sum of lump-sum taxes to consumers in the two countries $(T_{c1} + T_{c2})$ is negative. The third best differs from the second best because at the second best one of the countries, namely country 2, receives a transfer from the IEA lower than the carbon tax it has paid: $T_2 = 0.9 < \tau X_2 = 4.3$.

Country 1, which derives a higher utility from public spending than country 2, must levy a higher lump-sum tax on its consumers at the first best $(T_{c1} = -8.6 < T_{c2} = -5.5)$, receives higher transfers from the IEA $(T_1 > T_2)$, and consumes more public goods $(G_1 > G_2)$. At the

TABLE 5.3
First Best, Second Best, and Third Best for Equal Weights When Country 1 Values Public Spending More

	X_1	G_1	$T_{c1}\,|\,\psi_1$	ϕ_1	$\tau\,|\,\tau_1$	T_1	μ_1	U^1
				Country 1				
1st best	28.6	24.9	−8.6	0.085	0.372	13.9	0	447.6
2nd best	27.0	17.7	0.317	0.053	0.158	7.7	0.33	446.5
3rd best	25.2	15.6	0.435	0.038	0.149	3.7	0.47	441.3

	X_2	G_2	$T_{c2}\,|\,\psi_2$	ϕ_2	$\tau\,|\,\tau_2$	T_2	μ_2	U^2
				Country 2				
1st best	28.6	15.3	−5.5	0.085	0.372	7.4	0	393.1
2nd best	27.0	10.9	0.317	0.053	0.158	0.9	0.33	392.1
3rd best	29.1	12.7	0.172	0.077	0.187	5.5	0.17	396.8

second best, the marginal costs of public funds of the two countries are the same. The IEA then gives a large transfer to country 1 in order to allow it to consume more public goods. This result highlights the fact that the IEA may give large transfers to one country not only for equity motives but also because of the characteristics of preferences. At the third best, intercountry transfers do not take place and each country receives from the IEA a transfer exactly equal to the carbon tax it had paid. Without transfers from country 2, country 1's marginal cost of public funds increases a lot as it must resort to a high level of distortionary taxation to finance its public spending. Country 2 benefits (in terms of welfare) of the absence of intercountry transfers, whereas country 1's welfare is lower than at the second best.

3. Conclusion

Preexisting energy taxation and the presence of public goods that have to be financed do not necessarily imply to abandon the idea of a uniform carbon tax. However, the implementation of international transfers, for equity or other reasons linked to historical responsibilities or preferences, is required. If for various political economy reasons these intercountry transfers prove to be impossible, the carbon tax must be differentiated across countries. But are we sure that in the difficult context of international negotiations on climate change, making countries agree on different carbon taxes would be easier than making countries agree on a uniform tax and a system of transfers? In any case, we need a device promoting cooperation among countries and incentivizing participation to a climate agreement, along the line of the recent proposals by, among others, Nordhaus (2015), Weitzman (2015), or Cramton, Ockenfels, and Stoft (2015).

References

Atkinson, A. B., and J. E. Stiglitz. 1976. "The Design of Tax Structure. Direct Versus Indirect Taxation." *Journal of Public Economics* 6: 55–75.

d'Autume, A., K. Schubert, and C. Withagen. 2016. "Should the Carbon Price Be the Same in All Countries?" *Journal of Public Economic Theory* 18(5): 709–725.

Bergstrom, T. C. 1982. "On Capturing Oil Rents with a National Excise Tax." *American Economic Review* 72: 194–201.

Chichilnisky, G., and G. Heal. 1994. "Who Should Abate Carbon Emissions? An International Perspective." *Economic Letters* 44: 443–449.

Cramton, P., A. Ockenfels, and S. Stoft. 2015. "An International Carbon-Price Commitment Promotes Cooperation." *Economics of Energy and Environmental Policy* 4(2): 51–64.

Newbery, D. M. 2005. "Why Tax Energy? Towards a More Rational Policy." *The Energy Journal* 26(3): 1–39.

Nordhaus, W. 2015. "Climate Clubs: Overcoming Free-Riding in International Climate Policy." *American Economic Review* 105(4): 1339–1370.

OECD (Organization for Economic Co-operation and Development). 2015. *Taxing Energy Use 2015: OECD and Selected Partners Economies.* Paris: OECD Publishing.

Pigou, A. C. 1920. *The Economics of Welfare.* London: Macmillan.

Ramsey, F. 1927. "A Contribution to the Theory of Taxation." *Economic Journal* 35(145): 47–61.

Sandmo, A. 1975. "Optimal Taxation in the Presence of Externalities." *Swedish Journal of Economics* 77: 86–98.

Sandmo, A. 2004. "Environmental Taxation and Revenue for Development. In *New Sources of Development Finance*, edited by A. B. Atkinson. Oxford: Oxford University Press.

Weitzman, M. L. 2015. "Internalizing the Climate Externality: Can a Uniform Price Commitment Help?" *Economics of Energy and Environmental Policy* 4(2): 37–50.

CHAPTER SIX

Needed: Robustness in Climate Economics

TED LOCH-TEMZELIDES

THE PARIS COP-21 climate summit generated a fair amount of excitement and captured the public's imagination regarding future climate actions. The ratification of the Paris climate deal by the two biggest greenhouse gas (GHG) emitters, China and the United States, created anticipation about additional developments. This anticipation has been tempered by the recent decision by the current U.S. administration to abandon the Paris Accord. Despite this setback, both local and state governments as well as businesses in the United States appear committed to the Paris Accord. There is also hope that more nations will join in coordinating action in the future. Indeed, such a turn of events might well be considered necessary. Climate change affects the entire planet and, by definition, it requires global, coordinated, and self-enforcing solutions.

By now, it is all but certain that anthropogenic GHGs have an important effect on the earth's climate. It is also virtually certain that the resulting climate change will adversely affect economic activity and well-being in several regions around the globe. As fossil fuel use contributes to GHG emissions, an important question concerns the optimal future path for the fossil fuel and the renewable energy mix, or, in other words, the optimal energy transition. There is an important trade-off that we need to consider. GHG emissions aside, fossil fuels have supplied us with a cost effective way to fuel economic growth that has met the planet's energy needs since the Industrial Revolution. While exciting ongoing technological innovations are constantly reducing the costs of producing renewable energy, numerous innovations in fossil fuel extraction are also having a corresponding effect. Rigorous economic analysis is needed to balance the corresponding costs and benefits, and to prescribe the most desirable path among all available alternatives for our planet's current and future energy mix.

It is important to point out that anthropogenic climate change is unprecedented, and it involves substantial uncertainties. While the terms *risk* and *uncertainty* are used synonymously in everyday language, they indicate something very different to economists. Risk refers to situations where an outcome is random and, at the same time, we have sufficient experience with similar situations to have established reasonable probability assessments for all conceivable scenarios. Calculating probabilities of certain outcomes when rolling dice falls into this category. Uncertainty, on the other hand, refers to unfamiliar situations, in which there is randomness involved, but we have not had enough experience with the system in order to be able to form reliable probability assessments.[1]

I believe that this is exactly the situation we face when it comes to climate change. As I argue below, in order to give reliable policy advice, our climate economic models will need to be amended in order to deal with this kind of "deep uncertainty."

In his Nobel lecture, Hansen (2013) emphasized the need to consider uncertainty more broadly in economic modeling. As our models involve several simplifications, economists are always aware of the possibility of misspecification. This is a kind of *outside uncertainty*, as it represents the economist's concern about the model under study. However, we typically refrain from endowing the agents in the model with the same concern. This creates a discrepancy. Should we not allow the consumers, investors, and firms whose behavior we model to share the same concern as the economist? In short, putting the modeler and the agents in the model on the same footing requires that economic agents share the modeler's concern about mis-specification (inside uncertainty). How should economic agents respond to this concern when they make decisions?

In the study of economic models that incorporate climate change considerations, we must identify optimal behavior that takes into consideration the uncertainty associated with our limited knowledge about climate and its consequences. There is a well-established body of decision theory that suggests that, in the presence of such uncertainty, agents ought to optimize paying particular attention to worst-case scenarios.[2]

In the context of intertemporal decision making, this "maxmin" behavior can be operationalized using techniques from the field of *robust control*, a branch of control theory that studies optimal behavior that is tolerant to certain changes in the system under study. While not directly applicable to economics,[3] there is by now a well-established body of knowledge in control theory that involves optimizing the behavior of a system under an appropriate worst-case scenario. Examples include optimizing the stability of an airplane wing under maximal turbulence, or optimizing the structural properties of a building under the strongest earthquake within a class.[4] The basic idea behind this approach is that, since we are unsure of what the *true model* is, we should seek decision rules that perform well across several models in the neighborhood of our reference model. This desirable property is assumed to hold if the decision rules work well in the worst-case scenario model in this neighborhood.[5] An advantage of this approach is that it does not rely on any particular probabilistic structure. It only requires good performance across models that are considered close enough to be hard to distinguish given the observed data.

But what makes climate change special? Given the novelty of the problem, the corresponding uncertainty reflects our incomplete knowledge about both climate science and the economic consequences of climate change. The following example illustrates some of the complexities involved. We know that due to climate change, higher ocean water temperatures make it *more likely* that we will witness an increased frequency of hurricanes, say in America's Gulf Coast. We also know that, as they harvest energy from water temperatures, hurricanes are also likely to be more severe. However, when it comes to their actual formation and their characteristics, including location, frequency, severity, etc., and to calculating the value of the resulting economic damages, our assessments are, at best, tentative. This example is not special. It is fair to say that our understanding of most economic effects of GHG emissions is, at best, incomplete. In such situations, it is reasonable to prescribe an extra level of caution. More precisely, instead of optimizing taking into consideration past averages, we might be better off identifying and focusing on reasonable worst-case scenarios.

1. Climate-Economic Models Under Uncertainty

There is a small existing body of literature that attempts to incorporate concerns about model uncertainty into climate modeling. Hennlock (2009) introduces such uncertainty motivated by Ellsberg's (1961) paradox. He studies a two-sector integrated assessment model with endogenous growth. He identifies robust climate policy feedback rules that work well over a range of values. In the extreme case of "perfect uncertainty aversion," a zero-carbon consumption path can be desirable. Of course, perfect uncertainty aversion is inconsistent with other aspects of observed economic behavior. Among other reasons, it would be extremely costly to completely decarbonize the economy instantly, or even in the short run. Hennlock applies robust control under a more realistic worst-case scenario, using a two-sector model containing a carbon-intensive and a carbon-neutral sector. He proposes that the worst-case scenario is such that it is difficult to statistically distinguish it from reference models using detection error probabilities. He finds that model uncertainty aversion favors policies that avoid a realization of low probability, high negative impact outcomes. Assuming linear climate-related damages, uncertainty aversion increases carbon costs in a similar way as a low rate of time preference would. However, when damages are assumed to be nonlinear, desirable policy is more responsive, and it does not correspond to one generated by a calibrated low rate of time preference.

Funke and Paetz (2011) introduce model uncertainty in a linear quadratic model of a policy maker whose objective is to stabilize the atmospheric stock of carbon. Using robust control, they demonstrate that, under model uncertainty, the policy maker reacts more aggressively to changes in the stock of carbon in the atmosphere. They argue that emissions should be stabilized within the next five to seven years, independent of the underlying stabilization scenario.

Brock and Xepapadeas (2015) consider climate change in the context of an innovative model that considers the coevolution of the economy and the earth's ecosystem. The state of this coupled model involves ecological variables, such as species biomasses and the stock of greenhouse gases, as well as economic variables such as capital. They use robust control to model the structural uncertainty associated with climate change.

In the remainder of this section, I will discuss in more detail a recent model developed jointly by Li, Narajabad, and Temzelides (2016). The model extends the work of Golosov, Hassler, Krusell, and Tsyvinski (GHKT) (2013),[6] and contains the following components. First, model uncertainty is incorporated in a growth model with an energy sector. An environmental externality associated with GHG emissions is created by the use of energy in the production of goods. Second, the model considers "fat-tailed" distributions for damages. As Stern (2013), Weitzman (2014), and others have argued, when calculating expected costs from climate change, current models largely ignore "catastrophic" scenarios. Fat-tailed distributions partially address this concern by considering damage distributions under which "large" negative events occur with higher frequency than, say, under a normal distribution. Third, the model departs from the existing literature by considering unconventional fossil fuel sources. More precisely, it considers unconventional oil and gas, as well as methane hydrates in some of its simulations.[7] The paper uses robust control to characterize the optimal energy mix, carbon tax, and level of economic growth; i.e., the optimal energy transition, as a function of model uncertainty about the adverse effects of climate change.

More specifically, the model involves a representative agent maximizing the infinite sum of discounted period utility from final goods consumption. The final goods sector uses energy, capital, and labor, to produce output. Energy can come from three different sources: oil/gas, coal, and green energy. The use of oil/gas and coal creates GHG emissions, which subsequently enter the atmosphere. The carbon cycle is modeled as in GHKT. Model uncertainty about climate change is introduced through a stochastic variable which reduces the end-of-period capital stock. To use the earlier example, increased emissions can increase the severity and frequency of hurricanes which, in turn, can damage the economy's capital, say through flooding in certain areas. As is standard in robust control, the paper uses a Lagrange multiplier theorem to transform the concern about model uncertainty into a two-person dynamic game in which, after observing the representative agent's choice, a "malevolent player" chooses the worst-case specification of the model in each period. The malevolent player's deviation from the reference distribution is penalized by adding a term to the planner's objective function. This term increases in the distance between the reference distribution and the malevolent player's distribution choice.[8]

The problem of optimal growth involves maximizing lifetime utility for the representative agent, subject to the resource constraints and the evolution of GHGs. The main findings can be summarized as follows. By imposing the optimal (Pigouvian) tax associated with the GHG externality and then rebating the proceeds as lump-sum payments to consumers, the externality is completely internalized, and the resulting equilibrium allocation is efficient.[9] In other words, instead of subsidizing fossil fuel, a common practice around the world, fossil fuel should instead be taxed. The corresponding tax rates by fuel should reflect the marginal damage caused by the GHG emitted. As a practical example of a policy in that spirit, Michael Greenstone has suggested that mining leases should be adjusted to reflect the full climate damage from the corresponding fuels. Market forces would then lead to fossil fuels having the highest value (net of climate impact) being exploited first.[10] Indeed, some of the dirtiest fuels might well stay in the ground under such a scheme. Taking into account both economic growth and environmental externalities, the model prescribes a significant reduction in the use of coal, and a significant gradual increase in the production of renewable energy. Finally, it implies that, given its relative abundance and relative environmental friendliness, cleaner fossil fuel, such as natural gas, will be used intensely and for a long period of time in the transition, before renewables take over completely.

Perhaps more importantly, the analysis asserts that in the presence of uncertainty, not acting now can lead to dramatic consequences in the future. In a sense, active policies that promote renewables and natural gas relative to dirtier fuel such as coal, constitute a form of insurance against the uncertain consequences of climate change. While this insurance involves economic costs, as it implies a slowdown of world economic growth, it might well be worthwhile.

In the calibrated version of the model, the concern about uncertainty causes a significant decline in the optimal use of coal, and a smoother path for oil/gas consumption during the optimal energy transition. Absent a concern about model uncertainty, carbon concentration (net of the pre-industrial level) would rise almost threefold over the course of 200 years and reach a level of 600 Gt. In contrast, when concerns about model uncertainty are incorporated, the robust control approach implies that concentrations would increase by less than one-third and remain well below 300 Gt. The corresponding effect of model uncertainty on global temperatures is substantial. In the calibrated version of the model, carbon concentration under no concern about model uncertainty implies an average global temperature increase of more than 1.6°C over the course of 200 years, with global temperatures reaching more than 3°C above the pre-industrial level. In contrast, using robust control, the average global temperature increases by about 0.2°C. It is worth providing some perspective on the effects from various temperature increases. According to the Intergovernmental Panel on Climate Change (IPCC 2014), a 1°C increase would increase the risk of extreme weather events such as heat waves, heavy precipitation, and coastal flooding. Under a warming of above 2°C, the risks of affecting crop yields and water availability are perceived to be "high." Risks to the global economy are "moderate" under an additional 1-2°C warming, while few quantitative estimates are available for a warming of above 3°C. Risks associated with tipping points are perceived to be moderate between 0°C and 1°C, but increase under an additional 1-2°C warming, and are perceived to be high above 3°C.

2. Conclusion

The unprecedented nature of climate change creates the need to consider model uncertainty in economic models that incorporate a feedback between the world economy and the environment. This, in turn, motivates the use of robust control optimization. This novel approach has several implications for optimal growth, emissions, the energy transition, and for economic policy. A more sizable Pigouvian tax on emissions is needed to support the socially efficient allocation. This is in sharp contrast with existing policies, which subsidize fossil fuel. The concern about model uncertainty adversely affects the use of the dirtier fossil fuel, such as coal. In the presence of the fundamental uncertainties associated with climate change, active policies that promote renewables and natural gas relative to coal are a form of insurance.

This review abstracted from learning from experience with climate change and from technological progress in the energy sector. The latter is an important limitation, as significant technological developments are taking place in both renewable and fossil fuel energy production. This is an important topic for future research.[11]

Notes

This chapter is based on a presentation given at the IMF-OCP-Columbia University Seminar on "The Energy Transition, NDCs, and the Post-COP21 Agenda," held in Marrakech, Morocco, September 8–9, 2016.

1. This is sometimes also referred to as "deep uncertainty," "ambiguity," or "Knightian uncertainty." See Knight (1921).
2. See, for example, Gilboa and Schmeidler (1989).
3. For example, control theorists usually do not consider discounting or risk in their modeling.
4. Examples in economics include Epstein and Wang (1994), Hansen and Sargent (2001, 2005, 2008, 2010) and Cogley, Colacito, Hansen, and Sargent (2008). Lemoine and Traeger (2011) consider tipping points and ambiguity aversion.
5. For a comprehensive introduction of robust control in economic modeling see Hansen and Sargent (2008). Williams (2008) offers a non-technical review.
6. Related models can be found in Stern (2007), Nordhaus (2008), and van der Ploeg and Withagen (2012a, 2012b), to name a few.
7. Current estimates of methane hydrate resources are huge. See, for example, Rogner (1997), and Boswell and Collett (2011). While these resources are not economically recoverable with today's technologies and prices, this could well change in the coming decades, especially if demand strengthens.
8. See Basar and Bernhard (1995).
9. While other policy instruments can be considered, they would not improve over the outcome under the carbon tax, as this restores full efficiency in our model.
10. See https://epic.uchicago.edu/news-events/news/theres-formula-deciding-when-extract-fossil-fuels.
11. See, for example, Acemoglu et al (2012), and Adao, Narajabad, and Temzelides (2012).

References

Acemoglu, D., P. Aghion, L. Bursztyn, and D. Hemous. 2012. "The Environment and Directed Technical Change." *American Economic Review* 102 (1): 131–166.

Adao, B., B. Narajabad, and T. Temzelides. 2012. "A Model with Spillovers in the Adaptation of New Renewable Technologies." James A. Baker III Institute for Public Policy Working Paper.

Basar, T., and P. Bernhard. 1995. *H1-Optimal Control and Related Minimax Design Problems*. Boston: Birkhauser.

Boswell, R., and T.S. Collett. 2011. "Current Perspectives on Gas Hydrate Resources." *Energy and Environmental Science*, no. 4. 1206–1215.

Brock, W., and A. Xepapadeas. 2015. "Modeling Coupled Climate, Ecosystems, and Economic Systems." DEOS Working Papers 1508, Athens University of Economics and Business

Cogley, T., R. Colacito, L. P. Hansen, and T. J. Sargent. 2008. "Robustness and U.S. Monetary Policy Experimentation." *Journal of Money, Credit and Banking* 40 (8): 1599–1623.

Ellsberg, D. 1961. "Risk, Ambiguity and the Savage Axioms." *Quarterly Journal of Economics*, no. 75, 643–669.

Epstein, L. G., and T. Wang. 1994. "Intertemporal Asset Pricing under Knightian Uncertainty." *Econometrica* 62 (3): 283–322.

Funke, M., and M. Paetz. 2011. "Environmental Policy Under Model Uncertainty: A Robust Optimal Control Approach." *Climatic Change* 107 (3–4): 225–239.

Gilboa, I., and D. Schmeidler. 1989. "Maxmin Expected Utility with Non-Unique Prior." *Journal of Mathematical Economics* 18 (2): 141–153.

Golosov, M., J. Hassler, P. Krusell, and A. Tsyvinski. 2013. "Optimal Taxes on Fossil Fuel in General Equilibrium." *Econometrica* 82 (1): 41–88

Hansen, L. 2014. "Uncertainty Outside and Inside Economic Models." *Journal of Political Economy* 122 (5): 945–987.

Hansen, L., and T. Sargent. 2008. *Robustness*. Princeton, New Jersey: Princeton University Press.

Hansen, L. P., and T. J. Sargent. 2001. "Robust Control and Model Uncertainty." *American Economic Review* 91 (2): 60–66.

Hansen, L. P., and T. J. Sargent. 2005. "Robust Control and Model Misspecification." *Journal of Economic Theory* 128 (1): 45–90.

Hansen, L. P., and T. J. Sargent. 2010. "Wanting Robustness in Macroeconomics," in *Handbook of Monetary Economics*, vol. 3, edited by B. M. Friedman and M. Woodford (New York: Elsevier), 1097–1157.

Hennlock, M. 2009. "Robust Control in Global Warming Management. An Analytical Dynamic Integrated Assessment." Resources for the Future Discussion Paper, RFF DP 09-19, May.

IPCC (Intergovernmental Panel on Climate Change). 2014. Climate Change Synthesis Report, November.

Knight, F. H. 1921. *Risk, Uncertainty and Profit.* New York: Houghton Mifflin.

Lemoine, D. M., and C. Traeger. 2011. "Tipping Points and Ambiguity in the Economics of Climate Change." CUDARE Working Paper 1111R. Department of Agricultural and Resource Economics, University of California at Berkeley.

Li, X., B. Narajabad, and T. Temzelides. 2016. "Robust Dynamic Energy Use and Climate Change." *Quantitative Economics*, no. 7, 821–857.

Nordhaus, W. 2008. *A Question of Balance: Weighing the Options on Global Warming Policies.* New Haven, Connecticut: Yale University Press.

Nordhaus, W., and J. Boyer. 2000. *Warming the World: Economic Modeling of Global Warming.* Cambridge, Massachusetts: MIT Press.

Rogner, H.-H. 1997. "An Assessment of World Hydrocarbon Resources." *Annual Review of Energy and the Environment*, no. 22, 217–262.

Stern, N. 2007. *The Economics of Climate Change: The Stern Review.* Cambridge, UK/New York: Cambridge University Press.

Stern, N. 2013. "The Structure of Economic Modeling of the Potential Impacts of Climate Change: Grafting Gross Underestimation of Risk onto Already Narrow Science Models." *Journal of Economic Literature* 51 (3): 838–859.

van der Ploeg, F., and C. Withagen. 2012a. "Too Much Coal, Too Little Oil." *Journal of Public Economics* 96 (1–2): 62–77.

van der Ploeg, F., and C. Withagen. 2014. "Growth, Renewables and the Optimal Carbon Tax." *International Economic Review* 55 (1): 283–311.

Weitzman, M. L. 2014. "Fat Tails and the Social Cost of Carbon." *American Economic Review* 104 (5): 544–546. Williams, Noah. 2008. "Robust Control." In *The New Palgrave Dictionary of Economics*, 2nd ed., edited by Steven N. Durlauf and Lawrence E. Blume. London: Palgrave Macmillan.

PART THREE

Implementing Climate Agreements

CHAPTER SEVEN

Improving Paris

Credibility, Technology, and Conservation

BÅRD HARSTAD

Introduction

The United Nations approach to climate policy is and has been to focus on emission caps. Since emission of greenhouse gases is the direct cause of human-made climate change, capping country-specific emissions may at first appear to be a precise policy instrument. Thus, both the Kyoto Protocol and now the Paris Agreement aim at limiting country-specific emissions of greenhouse gases (UNFCCC 2016).

At the same time, it is well recognized that to be able to deal with reduced consumption of fossil fuels, the world must develop alternatives. Green technologies, such as abatement technologies and renewable energy sources, are necessary if we are to sustain our way of life. Despite this fact, the treaties have not discussed or attempted to pin down technology investments.

Ideally, one would think that technology will be driven by demand if countries and companies are limited in their capacities to emit. With such limits, the demand for technology will increase and firms will find it profitable to develop, invest in, and purchase new and green technology.

This chapter discusses these mechanisms and argues that one may need to question whether capping emissions will lead to the right development of new and green technology. After all, countries may act *strategically* when they contemplate how to invest or support the private sector's investments in technology. To account for, and to even take advantage of these strategic concerns, the climate treaty must be carefully designed, for example regarding how it is reviewed and revised over time.

Similarly, existing climate treaties, from Kyoto to Paris, have not attempted to regulate (or even discuss) the *supply side*. Since demand equals supply in a global market, one may hope and expect that regulating the demand for fossil fuels will necessarily lead to less fossil fuels being extracted from the ground. In some ideal settings, regulating only the demand side can be sufficient. In the real world, however, there are several reasons why it may be both beneficial and necessary to consider an additional regulation of the supply side, and thus the countries' extraction levels of fossil fuels. This chapter will be quite precise and concrete on how to go ahead with such supply-side regulation.

Paying countries to reduce deforestation in the tropics is also a type of supply-side environmental policy. In contrast, a demand-side policy would be to boycott timber or agricultural products produced on land converted from forests, but such a boycott is likely to reduce the price for other buyers, who will then purchase more. This type of *leakage* can be avoided by a well-designed supply-side environmental policy.

As discussed in the final section, some of the discussed reasoning is illustrated by the 2017 notification to the UN that the United States intends to withdraw from the Paris Agreement. At the same time, this possibility strengthens the below arguments for focusing on technology and the supply side, and challenges our views on whether one needs to view climate agreements and international trade together.

1. Treaties, Technology, and the 2016 Nobel Prize in Economics

Technology requires investments. By definition, investments are costs paid today in return for benefits tomorrow. It is thus essential that one anticipates that the technology is likely to be useful later, before private companies can be expected to invest, and also before individual countries are willing to support investments in environmentally friendly technology. By *green* technology, I am referring to any type of environmentally friendly technology, such as renewable energy sources, abatement technology, or even carbon capture and storage (CCS) technologies.

Naturally, private investors will invest more in green technology if countries have credibly promised to emit less greenhouse gases in the future. However, as long as future commitments are *expected*, companies will invest more today in order to maximize profits. No *guarantees* are necessary to motivate firms to invest.

Instead, the main problem arises when countries negotiate and quantify emission levels without at the same time negotiating how much they should invest and thus tie their hands to a sustainable path going forward. The problem is that once a country has invested heavily and succeeded with a green transition in its economy, this country has less clout in future climate negotiations since it is no longer a threat to other countries fearing emissions from the others. In a typical bargaining game, such a country can be requested to cut emissions by more since that would be cost effective, and, in fact, also *fair* in an ex post perspective where one perceives past investments as sunk. When one anticipates that today's investments will be met by larger demands tomorrow, the incentive to invest is naturally diminished.

This underinvestment problem is the so-called *hold-up problem* in economics. The hold-up problem is a central issue in the research of Oliver Hart, the 2016 Nobel Prize winner in Economics (Grossman and Hart 1986). Hart's research, and the subsequent literature, has shed light on how the hold-up problem can be managed and dealt with by setting up clever agreements between the negotiators.

Climate negotiators can learn from this insight. In particular, the problem of underinvestment will be smaller, and investments will be larger, if the duration of the commitment period is longer. If investments in green technology are believed to be important, then the duration of the commitment period should be longer (Harstad 2016a).

2. Reviewing the Review Mechanism

The review mechanism in the Paris Agreement states that countries should revise and update their commitments every five years. It is necessary to ratchet up the commitments over time as we improve technology and thus our ability to make larger cuts. It has not been specified exactly how the revised commitments are to be decided or negotiated, however, and here the *devil's in the details*, it turns out.

In particular, suppose that countries (re)negotiate their commitments every five years under the presumption that if these negotiations halt and break down, then one moves forward without any commitments at all. For this type of bargaining game, it is the technology laggards that will be rewarded with larger emission allowances in the equilibrium bargaining outcome, because giving the laggards larger quotas will be necessary to secure their continued support. Anticipating this, countries will be reluctant to invest up front, exactly as described in the previous section, where the hold-up problem was explained.

If instead, the default outcome of the (re)negotiation is the past set of emission cuts, or commitments, then there is a limit as to how much the laggards can expect to gain by holding back on technology investments. In fact, the countries that have already invested and prepared for the previously-agreed-to emission cuts will be comfortable with the status quo, and thus they will be in a good bargaining position. In order to obtain that good bargaining position, countries will invest more than in the situation in which the default outcome were no commitment at all.

Note that trade liberalization talks are organized in the latter way. Trade liberalization has occurred over multiple trade talks in recent decades. The commitments to liberalization have (rarely) been earmarked with expiration dates, and the presumption has always been that if one new round of trade talks fails, one returns to the existing set of trade agreements rather than to autarky. This way of organizing negotiations is more efficient, according to the reasoning explained above. In other words, climate negotiators can learn from successful trade liberalization negotiations.

In fact, an even better system for climate treaties is to return to an ever more ambitious plan for emission cuts, if later (re)negotiations should happen to fail. That is, the best treaty design is to negotiate and commit to a long-term path in which emission caps are forever decreasing, even though countries periodically (for example, every five years) may return to these commitments and revise or renegotiate them, depending on the circumstances. As long as the negotiators expect that the default outcome is not the business-as-usual scenario, but instead a path with decreasing emission caps, then they will not have to reward countries that have underinvested when they (re)negotiate new caps. As a consequence, countries will invest more (Harstad 2012a).

3. Credibility in the International Prisoner Dilemma Game

International treaties cannot easily be enforced by third parties or harsh penalties. At most, a country that emits more than it promised may expect to lose goodwill or the other countries' willingness to comply with their parts of the treaty. Therefore, international climate treaties must be self-enforcing: Complying with the treaty must be in the best interest of the country.

One problem with a self-enforcing agreement is that climate change and environmental problems can be viewed as a prisoner dilemma game. That is, in a one-shot setting, it may be in the interest of a single country to emit rather than abate, but all countries may be worse off when everyone follows this strategy. A better outcome for everyone would be that every country abates, but that may not be individually rational when each country takes as given the actions of the others.

Cooperation in a prisoner dilemma game can be sustained when the game is repeated and countries worry that if one cheats by emitting today, other countries will do the same in the future. For cooperation and abatement to be individually rational in this context, two conditions must be satisfied. First, countries must care sufficiently about the future. Second, the temptation to *cheat* and emit rather than abate cannot be too large.

The gains from cheating and emitting rather than abating depend on a country's stocks of "green" and "brown" technologies. If a country has renewable energy sources, or if it is able to clean emissions effectively, then the cost of abating relative to emitting is smaller, and the temptation to cheat in the prisoner dilemma game is diminished. If instead, a country is endowed with a *brown* industry structure, as after advancements in the extraction of (unconventional) fossil fuels, then the country will be more tempted to emit. These perspectives may shed some light on President Trump's announcement in 2017 that he seeks to withdraw the United States from the Paris Agreement.

When other countries find it credible that a country with more green than brown technology is likely to cooperate by abating, then these other countries may also be willing to comply with their promises to prevent cooperation from breaking down. Thus, to raise the likelihood for compliance and ensure that the treaty is self-enforcing, it may be necessary to require countries to invest more in green technology and correspondingly less in brown technology (Harstad, Lancia, and Russo 2017).

It is evidently complicated to require a country such as the United States to invest more in green and less in brown technologies. There are few but some alternative ways in which one can raise credibility and the countries' willingness to cooperate. One method is to introduce sanctions on countries that do not comply or participate. Sanctions can make a climate treaty self-enforcing; even if no sanction will ever need to be imposed in practice (because, when the sanction is credible, countries will find it in their self-interest to comply). The term *sanction* has a negative connotation, but such a mechanism can be framed positively by stating that a country that complies will be granted the so-called *most favored nation* status when it comes to international trade, implying larger market access and/or lower tariffs or nontariff border measures.

Another way to motivate compliance and participation is to reduce the benefits of free riding by influencing the (global) supply of fossil fuels. This mechanism is discussed in the next section.

4. The Supply Side: Regulate Fossil Fuel Extraction?

Demand equals supply in a global market. Thus, the sum of countries' fossil fuel consumption equals the amount that is extracted from the ground or the sea. If one side of the market is regulated, then the other side of the market will adjust accordingly. In other words, if countries cap their consumption of fossil fuels, then fossil fuel producers will face lower demand and they will find it profitable to reduce extraction by the same amount. Nevertheless, it can be highly beneficial to keep an eye on the supply side as well, for several reasons.

1. Since regulating the consumption of fossil fuels will reduce demand and thus the global fossil fuel price, the temptation to cheat by exploiting the low price and consuming more fossil fuel becomes more tempting. As explained in section 3, an agreement

on reducing emissions is more likely to be self-enforcing if the temptation to emit more is reduced. The temptation is reduced if the price of fossil fuel is high. The price is high if one seeks to reduce global supply (by requiring countries to extract less) in addition to reducing consumption and demand.

2. Relatedly, the higher fossil fuel price that follows if extraction is regulated will also reduce the countries' willingness to free ride by not participating in the treaty in the first place. For example, importers of fossil fuel will prefer that their country does not participate in a climate treaty if only consumption is regulated, since the associated lower price makes it particularly profitable to remain a nonparticipant. Therefore, the pressure from lobbyists to not sign/ratify a treaty can be larger if the focus is exclusively on regulating end-of-the-pipe emissions.

3. Even countries that do end up free riding will find it beneficial to invest in green technology, if the fossil fuel price is also high for them (i.e., if extraction is reduced); see Harstad (2012b).

4. Since an exclusive focus on regulating consumption will lead to a low global price, fossil fuel producers and exporting countries will be severely harmed by such a climate treaty, and they will work against it. To secure their support and participation, and to end up with a fair outcome in which the price is stabilized, one may want to regulate countries' extraction levels in addition to their emissions.

5. Regulating both sides of the market works as a global insurance in the risk of a failed treaty. If the Paris Agreement succeeds, so that the consumption of fossil fuels is reduced, then extraction will also be reduced, as explained at the beginning of this section. In this light, an additional regulation of extraction will have little impact on efficiency (except that it will lead to a larger fossil fuel price, as discussed above). However, if the Paris Agreement happens to fail, in the sense that countries end up consuming and emitting more than they pledge, or if other countries follow President Trump's lead in withdrawing from the agreement, then an additional agreement on reducing extraction levels may ensure that global emissions cannot elevate to a very large level. With this perspective, an additional agreement on extraction will have no downside, but the upside is that it can function as an insurance, if the demand-side policy works less effectively than we hope.

5. A Moratorium on Arctic Resources?

Given the benefit of regulating fossil fuel extraction, one may argue that one geographical area that is particularly suitable for such a regulation is the Arctic. There are several justifications for this claim.

1. Fossil fuel resources in the Arctic will be costly to extract, and they are thus not among the most profitable natural resources, neither from a private nor from a public perspective.
2. These resources are also environmentally risky to extract, since the ecological system in cold water is particularly sensitive to oil spills.
3. The technology required for an effective exploration in the Arctic has yet to be developed. The cost of this investment can be avoided if one abandons the plans for drilling in the Arctic.
4. The property rights or extraction rights in the Arctic Sea are claimed by several countries. Thus, multiple countries will bear the burden of not being able to extract in the high North. These countries are also relatively rich, so they should be able to bear this burden without making it necessary to compensate them with explicit side transfers. In fact, limiting extraction will lead to a larger fossil fuel price, as explained above, and the larger price is beneficial to fossil fuel exporters.
5. The claims to property rights in the Arctic Sea are to some extent overlapping. A moratorium will reduce the tension and the potential for conflict over these resources.
6. As of 2018, the world is still uncertain about (i) who will be able to secure which property rights in the Arctic Sea, (ii) the exact values or locations of the resources that can be extracted, and (iii) the cost and effectiveness of the technology that one needs in the process. These uncertainties imply that the countries with claims are all *behind the veil of ignorance*, to some extent. That is, for these countries the expected benefits of exploration are more similar today than they will be in a decade or so, when some of the uncertainty is clarified. This similarity should make it easier to sign a moratorium for the Arctic today, than it will be later in the future. One thus has a unique possibility to negotiate a moratorium for the Arctic Sea today; an opportunity that may be lost in some years.

7. It is certainly challenging to negotiate a moratorium for Arctic resources. However, the Antarctic Treaty (banning military operations and later resource extraction in Antarctica) was negotiated and signed at the midst of the Cold War (1959–1962). So, an analogous treaty for the Arctic should be possible in our time, even when one recognizes that the Arctic is a sea, and that the *Laws of the Sea* are different from the laws that were present (or absent) for Antarctica.

6. Deforestation and Forest Conservation

The policy discussed in sections 4 and 5, on regulating fossil fuel extraction, is related to the policy of incentivizing reduced deforestation. To see this, a demand-side approach to reducing deforestation in the tropics would be to reduce consumption and to boycott timber or agricultural products from such areas. Such a boycott will reduce the price of such products, and buyers/countries that are not participating in the boycott will purchase more. Given such a leakage, it is more effective to conserve particular areas of forests and regulate extractions of these resources directly (in other words, focus on the supply side of the market).

An important difference to fossil fuel resources is that the countries owning tropical forests are few and relatively poor. Thus, it is clearly necessary to offer explicit compensation to them in exchange for conserving their forests.

Such a policy is extremely cost effective, according to a number of studies. Not only is a large share of global greenhouse gas emissions coming from deforestation in the tropics, but deforestation also leads to huge losses of biodiversity and the homes of the world's last indigenous tribes. The global costs of deforestation amount to $2–$4.5 trillion a year, according to *The Economist* (2010). At the same time, estimates suggest that deforestation can be halved at a cost of $21–$35 billion per year, or reduced by 20–30 percent at a price of $10/$tCO_2$ (IPCC 2014; Busch et al. 2012).

To succeed with forest conservation, it is urgently needed to announce and promise funds to be used for compensations. If countries in the tropics fear that demand for their products will be reduced in the future, they become motivated to log and remove forests today (this is related to the so-called "Green Paradox" [Sinn 2008]). If instead, one can credibly expect to be compensated for conserving forests in the future, then countries will be motivated to conserve today, even if the funds are expected to be released only at some point in time in the future (Harstad 2016b).

7. When Countries Withdraw or Fail to Comply

In August, 2017, U.S. President Trump notified the United Nations about the White House's intention to withdraw from the Paris Agreement. This decision followed the 2016 U.S. election in which the winner was a candidate that is often claimed to be unusually non-traditional and populistic. At the same time, the notification is consistent with much of the reasoning above: Only binding commitments to cut emissions may sufficiently motivate (discourage) the development of green (brown) technology, and the United States did not face such commitments during the previous agreement, the Kyoto Protocol. Without a transition from brown industry structures, the temptation to emit rather than abate naturally dominated. Since there is no trade sanctions/carrots associated with participation, a large country faces few consequences when deciding to free ride.

The possibility to withdraw and free ride documents the challenges of relying on self-enforcing agreements. It is simply not credible that the defection of one country should trigger other countries to raise their emission levels, particularly since renegotiation is always possible in the real world. These possibilities strengthen the above conclusion that participants ought to invest more in green and less in brown technology for compliance to be credible. Withdrawals and free riding also show that it is necessary to accompany the regulation on demand with regulation of the extraction of fossil fuel and deforestation in order to limit carbon leakage.

While optimists may hope that the policies discussed above can be sufficient to motivate participation and participants, pessimists fear they are not. Since there is no world government in international politics, it is hard for progressive countries to motivate unwilling countries to participate. At the end of the day, the only instruments available in international politics may be to tie one type of agreement to another. Although the international community has thus far been reluctant to risk the world trading system as an enforcer of environmental agreements, the future will show whether such a bundling is necessary to motivate participation and compliance.

Note

This chapter is based on my presentation at the IMF-OCP Columbia University Seminar on "The Energy Transition, NDCs, and the Post-COP21 Agenda," held in Marrakech, Morocco, September 8–9, 2016.

References

Busch, J., R. N. Lubowski, F. Godoy, M. Steininger, A. A. Yusuf, K. Austin, J. Hewson, D. Juhn, M. Farid, and F. Boltz. 2012. "Structuring Economic Incentives to Reduce Emissions from Deforestation within Indonesia." *Proceedings of the National Academy of Sciences* 109 (4): 1062–1067.

The Economist. 2010. "Money Can Grow on Trees." Available at www.economist .com/node/17062651.

Grossman, S., and O. Hart. 1986. "The Costs and Benefits of Ownership: A Theory of Vertical and Lateral Integration." *Journal of Political Economy*, no. 94, 691–719.

Harstad, B. 2012a. "Climate Contracts: A Game of Emissions, Investments, Negotiations, and Renegotiations." *Review of Economic Studies* 79(4): 1527–1557.

Harstad, B. 2012b. "Buy Coal! A Case for Supply-Side Environmental Policy." *Journal of Political Economy* 120(1): 77–115.

Harstad, B. 2016a. "The Dynamics of Climate Agreements." *Journal of the European Economic Association* 14(3): 719–752.

Harstad, B. 2016b. "The Market for Conservation and Other Hostages." *Journal of Economic Theory*, no. 166, 124–151.

Harstad, B., F. Lancia, and A. Russo. 2017. "Compliance Technology and Self-Enforcing Agreements." CESifo Working Paper No. 5562. University of Oslo, Norway.

IPCC (Intergovernmental Panel on Climate Change). 2014. *Climate Change 2014: Mitigation of Climate Change.* Contribution of Working Group III to the Fifth Assessment Report of the Intergovernmental Panel on Climate Change (edited by Edenhofer, O., R. Pichs-Madruga, Y. Sokona, E. Farahani, S. Kadner, K. Seyboth, A. Adler, I. Baum, S. Brunner, P. Eickemeier, B. Kriemann, J. Savolainen, S. Schlömer, C. von Stechow,

T. Zwickel, and J.C. Minx). Cambridge, UK/New York: Cambridge University Press.

Sinn, Hans-Werner. 2008. "Public Policies Against Global Warming: A Supply Side Approach." *International Tax and Public Finance*, no. 15, 360–394.

UNFCCC (United Nations Framework Convention on Climate Change). 2016. "Report of the Conference of the Parties on Its Twenty-First Session." http://unfccc.int/resource/docs/2015/cop21/eng/10.pdf.

CHAPTER EIGHT

Can a Uniform Carbon-Price Commitment Help to Resolve the Global Warming Problem?

MARTIN L. WEITZMAN

Introduction

Throughout this chapter I use the terms *climate change* and *global warming* interchangeably. The term climate change is currently in vogue and is a more apt description overall. But the term global warming is more evocative of this chapter's main theme. Global warming is a *global* public goods externality whose resolution requires an unprecedented degree of international cooperation and coordination. This international climate change externality has frequently been characterized as the most difficult public goods problem that humanity has ever faced. I concentrate in this chapter on carbon dioxide emissions, but in principle the discussion could be extended to emissions of all relevant greenhouse gases. Throughout the chapter I blur the distinction between carbon dioxide and carbon, since the two are linearly related.[1]

The core problem confronting the political economy of climate change is an inability to overcome the obstacles associated with free riding on a very important international public good. The "international" part is significant. Even within a nation, it can be difficult to resolve public goods problems. But at least there is a national government, with some governance structure, able to exert some control over externalities within its borders. A national government can (at least in principle) *impose* targets on national public goods. With climate change there is no overarching international governance mechanism capable of coordinating the actions necessary to overcome the international problem of free riding. Instead, instruments of control, like prices and/or quantities, must be *negotiated* among sovereign nations.

My point of departure throughout all of what follows is the critical centrality of the international free-rider problem as a cause (really *the* cause) of negotiating difficulties on climate change emissions. Negotiators here are playing a game in which self-interested strategies are a crucial consideration. It turns out that negotiating rules define an important part of the game, and can thereby change self-interest, for better or for worse.

In this chapter I try to argue that a uniform minimum global tax-like price on carbon emissions, whose revenues each country retains, can provide a focal point for a reciprocal common commitment, while quantity targets, which do not as readily present such a single focal point, have a tendency to rely ultimately on individual commitments. As a consequence, negotiating a uniform minimum global carbon tax or price can help to solve the externality problem while individual caps essentially incorporate it. I will try to explain why negotiating a uniform minimum carbon price embodies what I call a "countervailing force" against narrow self-interest by automatically incentivizing all negotiating parties to internalize, at least approximately, the global warming externality. The basic challenge, as I see it, is to construct a relatively simple and relatively acceptable one-dimensional international *quid pro quo* mechanism, which automatically embodies the principle of "I will if you will."

Throughout this chapter I argue that it is very difficult to resolve the global warming externality problem by directly assigning individual quantity targets. A meaningful comprehensive quantity-based treaty involves specifying as many different binding emissions quotas (whether in the form of tradable permits or not) as there are national entities. Each national entity has a self-interested incentive to negotiate for itself a high cap on carbon emissions—much higher than would be socially optimal. The resulting free-rider problem plagues a quantity-based approach. Even if there were a collective commitment to negotiate or vote on a second-stage worldwide total emissions cap, disagreements over the first-stage fractional subdivision formula (for disaggregating the negotiated or majority voted aggregate worldwide quantity cap into individual quantity caps) would make it difficult to enact such a quantity-based approach.[2]

The inspiration for this chapter is the perception of a desperate need for some radical rethinking of international climate policy. As a possibly useful conceptual guide for what negotiations might accomplish, I sometimes ask the reader to temporarily suspend disbelief by considering what might happen in a World Climate Assembly (WCA) that votes on global carbon emissions via the basic principle of a one-person-one-vote majority rule. In this conceptualization, nations would vote along a single dimension for their desired level of emissions stringency on behalf of their citizen constituents, but the votes are weighted by each nation's population.

Right now, anything like a WCA seems hypothetical and hopelessly futuristic. It presumes a state of mind where the climate change problem has become sufficiently threatening on a grassroots level that world public opinion is ready to consider novel governance structures which involve relinquishing some national sovereignty in favor of the greater good. What might be the justification for a new international organization like the WCA? The ultimate justification is that big new problems may require big new solutions. For a world desperately wanting new solutions to the important externality of climate change, perhaps it is at least worth considering establishing a new organization along the lines of the WCA. After all, it is useful to have some concrete fallback decision mechanism behind vague "negotiations" because even with the focus on a one-dimensional harmonized carbon price (or with the focus on a one-dimensional quantity of total emissions), there are bound to be disagreements whose resolution is unclear. I merely assume that it is in the interest of enough nations to forfeit their rights to pollute in favor of a WCA voting solution of the global warming externality. This is truly a heroic assumption at the present time because the WCA does not correspond to any currently existing international body. Taken less literally, the thought experiment of a hypothetical WCA can still help us to concentrate our thinking and intuition on what negotiations should be trying to accomplish. In other words, I am hoping that the fiction of a WCA might be useful in indicating what might be the outcome of less-formal international negotiations.

It could be objected that a consensus voting rule, not a majority voting rule, is employed in negotiations under the United Nations Framework on Climate Change. This consensus voting rule has been widely interpreted as requiring near unanimity. With such a restrictive voting rule, significant progress on resolving the global warming externality seems virtually impossible. Surely, a less restrictive voting-like rule, such as majority rule, would render progress more likely, and is at least worth considering.

One aspect should perhaps be emphasized above all others at the outset. The global warming externality problem is highly unlikely to be resolved without a binding agreement on some overall formula for dividing emissions responsibilities among nations. Voluntary altruism alone is unlikely to solve this international public goods problem. Of necessity there must be some impingement on national sovereignty in the form of an international mechanism for determining targets, verifying fulfillment, and punishing noncompliance. The question then becomes: *Which* collective-commitment frameworks and formulas are more promising than which others?

1. Negotiating a Uniform Carbon Price

In this chapter I examine the theoretical properties of a natural one-dimensional focus on negotiating a single binding price on carbon emissions, the proceeds from which are domestically retained. As previously mentioned, for expositional simplicity, I identify this single binding price on carbon as if it is an internationally harmonized nationally collected carbon tax. At a theoretical level of abstraction, I blur the distinction between a carbon price and a carbon tax. However, in actuality the important thing is acquiescence by each nation to a common binding minimum price on carbon emissions, not the particular mechanism by which this common binding minimum price is attained by a particular nation.

A system of uniform national carbon taxes with revenues kept in the taxing country is a relatively simple and transparent way to achieve internationally harmonized carbon prices. But it is not necessary for the conclusions of this chapter. Nations or regions could meet the obligation of a minimum price on carbon emissions by whatever internal mechanism they choose—a tax, a cap-and-trade system, a hybrid system, or whatever else results in an observable price of carbon above the internationally agreed minimum. I elaborate further on this issue in my concluding remarks.

At a theoretical level, I would suggest that the instruments of negotiation for helping to resolve the global warming externality should ideally possess three desirable properties.

1. *Induce cost effectiveness.*
2. Be of *one-dimension based on a "natural" focal point* to facilitate finding an agreement with relatively low-transactions costs.
3. *Embody a "countervailing force" against narrow self-interest by automatically incentivizing all negotiating parties to internalize the externality via a simple, reciprocal,* I will if you will, *common climate–commitment formula.*

Using these three desirable theoretical properties as criteria, I now compare and contrast an idealized binding harmonized tax-like price with an idealized binding cap-and-trade system (without a uniform price floor).

On the first desirable property, in principle both a carbon price and tradable permits achieve cost effectiveness (provided agreement can be made in the first place).

The second desirable property (low dimensionality) argues in favor of a one-dimensional harmonized tax-like carbon price over an n-dimensional harmonized cap-and-trade system among n nations. Alas, this argument is elusively difficult to formulate rigorously, or even to articulate coherently. My argument here is necessarily intuitive or behavioral and relies on empirical counterexamples. In this situation, two important empirical counterexamples are the breakdown of the quantity-based Kyoto approach and the underambitious "intended nationally determined contributions" actually volunteered by nations under the COP21 Paris Accord.

With n different national entities, a quantity-based treaty involves assigning n different binding emissions quotas (whether tradable or not). Treaty making can be viewed as a coordination game with n different players. Such a game can have multiple solutions, often depending delicately on the setup, what is being assumed, and, most relevant here, the choice of negotiating instrument. In the case of Kyoto, the world had in practice arrived at a bad quantity-based solution that essentially devolved to regional volunteerism. The ultimate outcome of the COP21 Paris Accord remains to be seen, but so far the intended nationally determined contributions (INDCs) actually volunteered by the parties seem markedly underwhelming, even leaving aside the near impossibility of achieving the stated goal of keeping global warming below 2°C.

Thomas Schelling introduced and popularized the notion of a focal point in game theory.[3] Generally speaking, a focal point of an n-party coordination game is some salient feature that reduces the dimensionality of the problem and simplifies the negotiations by limiting bargaining by the parties to some manageable subset, hopefully of one dimension. The basic idea is that by limiting bargaining to a salient focus, there may be more hope of reaching a good outcome. In a somewhat circular definition, a focal point is anything that provides a focus of convergence. The "naturalness" or "salience" of a focal point is an important aspect of Schelling's argument that is difficult to define rigorously and is ultimately intuitive.

The concept of "transaction cost" is associated with the work of Ronald Coase.[4] The basic idea is that n parties to a negotiation can be prevented from attaining a socially desirable outcome by the costs of transacting the agreement among themselves. One could try to argue that, other things being equal, transaction costs increase proportionally with the number of parties n. Negotiating a one-dimensional price with single-peaked preferences has the important additional property of allowing a majority-rule voting equilibrium, which avoids the Arrow impossibility theorem.

In the case of international negotiations on climate change, I believe that both Schelling's concept of a salient focal point and Coase's concept of transactions costs can be used as informal arguments to support negotiating a single-harmonized carbon price whose proceeds are nationally rebated. Put directly, it is easier to negotiate one price than n quantities—especially when the one price can be interpreted as fair in terms of equality of marginal effort. I cannot defend this claim rigorously. At the end of the day, this is more of a plausible conjecture than a rigorous theorem. Whether justly or not, throughout this chapter I basically assume that the essential contrast is between one binding price assignment versus n binding quantity assignments—and I then proceed to examine the consequences.

The third desirable property is that the instrument or instruments of negotiation should embody a countervailing force against narrow free-riding self-interest by incorporating incentives that automatically internalize the externality. Such incentives should ideally take the form of a simple, reciprocal, common climate commitment based on the quid-pro-quo principle of *I will if you will.* I believe this third property is arguably the most important property of all. This countervailing force property is inherently built into a price-based harmonized system of emissions charges, but it is absent from a quantity-based international cap-and-trade system, at least as traditionally formulated.

If I am assigned a cap on emissions, then it is in my own narrow free-riding self-interest to want my cap to be as large as possible (whether or not my cap will be tradable as a permit). The self-interested part of me wants maximal leniency for myself. Other than altruism, there is no countervailing force on the other side encouraging me to lower my desired emissions cap because of the externality benefits I will be bestowing on others.

Within a nation, the government *assigns* binding caps. But *among* sovereign nations, binding caps must be *negotiated.* I believe that this is a crucial distinction for the success or failure of a cap-and-trade regime. A quantity-based international system fails because no one has an incentive to internalize the externality and everyone has the self-interested incentive to free ride. What remains is essentially an erratic pattern of altruistic individual volunteerism that is far from a socially optimal resolution of the problem.

An internationally harmonized domestically collected carbon price is different. If the price were imposed on me alone, I would wish it to be as low as possible so as to limit my abatement costs. But when the price is uniformly imposed, it embodies a countervailing force that internalizes the externality for me. Counterbalancing my desire for the price to be low (in order to limit my abatement costs) is my desire for the price to be high so that other nations will restrict their emissions, thereby increasing my benefit from worldwide total carbon abatement. A binding uniform minimum price of carbon emissions has a built-in self-enforcing mechanism that countervails free riding.[5]

In previous work, I have tried to model formally the role of this third "countervailing force" property of an internationally harmonized but nationally collected carbon price.[6] I constructed a basic model indicating an exact sense in which each agent's extra cost from a higher international minimum emissions price is counterbalanced by that agent's extra benefit from inducing all other agents to simultaneously lower their emissions via the higher international minimum price (which might well take the form of a uniform price floor on a cap-and-trade system).

With further restrictions, the model showed that population-weighted majority rule for an internationally harmonized tax-like carbon price can come as close to an optimal price on emissions as the median per-capita marginal benefit is close to the mean per-capita marginal benefit. The key insight from this way of looking at things is that in voting (or more generally negotiating) a universal minimum carbon price, various nations are, to a greater or lesser degree, internalizing the externality. Loosely speaking, an "average" nation is fully internalizing the externality because its extra cost from a higher emissions price is exactly offset by its extra benefit from inducing all other nations to simultaneously lower their emissions via the higher price.

On the price side, a uniform carbon price automatically has the desirable property that cost effectiveness is guaranteed. I think that the formal WCA voting result of the model might perhaps be interpreted somewhat less formally as indicating that negotiating an internationally harmonized (but nationally collected) carbon price may have an important desirable property on the quantity side as well. If the median marginal benefit (per capita) equals the mean marginal benefit (per capita), then the socially optimal carbon price in the model has the property that, roughly speaking, half of the world's population wants the price to be higher, while the other half of the world's population wants the price to be lower. In this situation, the desirable quantity-side property is that the total worldwide output of all emissions might be close to being optimal to the extent that the outcome of negotiations mimics the outcome of majority voting. Although the real world is a far more complicated and nuanced place than the restrictive theoretical model that was constructed, I think this voting result is trying to indicate something positive (even if only at an abstract level) about how a negotiated uniform carbon price might possess some overall potential to counteract via internalization the externality of global warming.

2. Concluding Remarks

At the end of the day, there is no airtight logic in favor of a negotiated price over negotiated quantities, only a series of partial arguments.

One argument is that the revenues from a tax-like carbon price are nationally collected, so that the contentious distributional side is somewhat hidden and there is at least the appearance of fairness as measured by equality of marginal effort. A second desirable feature, I have argued, is the natural salience and relatively low-transaction costs of negotiating one price as against negotiating n quantities, which, while somewhat imprecise, is in my opinion an important distinction. A third argument is the self-enforcement mechanism that constitutes the main theme of this chapter, namely the built-in countervailing force of an imposed uniform price of carbon that tends to internalize the externality and gives national negotiators an incentive to offset their natural impulse to otherwise bargain for a low price.

Of necessity, my argument has been sprinkled with subjective judgments. This, unfortunately, is the nature of the subject. To repeat yet again, this time after examining somewhat more carefully the alternatives, I judge it difficult to escape the conclusion that, in the context of an international treaty that covers all major emitters, it is more politically acceptable and it comes closer to a social optimum to negotiate one binding price than n-binding quantities or quantity-like distributional coefficients.

My argument here is sufficiently abstract that it is open to enormous amounts of criticism on many different levels. There are so many potential complaints that it would be incongruous to list them all and attempt to address them one by one. These many potential criticisms notwithstanding, I believe the argument here is exposing a fundamental countervailing-force argument that deserves to be highlighted.

Because the formulation is at such a high level of abstraction, it has blurred the distinction between a carbon price and a carbon tax. As previously noted, the important thing is acquiescence by each nation to a *binding minimum price* on carbon emissions, not the particular internal mechanism by which this obligation is met. A system of national carbon taxes with revenues kept in the taxing country is a relatively simple and transparent way to achieve internationally harmonized carbon prices. But it is not absolutely necessary for the conclusions of this chapter. In principle, nations or regions could meet the obligation of a minimum price on carbon emissions by whatever internal mechanism they choose—a tax, a cap-and-trade system with a tax floor, some other hybrid system, or whatever else results in an observable price of carbon above the uniform minimum.[7]

Of course any nation or region could choose to impose a carbon tax or price above the international minimum. The hope is that even a low positive initial value of a universal minimum carbon tax or price could be useful for gaining confidence and building trust in this price-based international architecture.

The purpose of this chapter is primarily expository and exploratory. *Any* proposal to resolve the global warming externality will face a seemingly overwhelming array of practical administrative obstacles and will need to overcome powerful vested interests. That is the nature of the global warming externality problem. The theory of this chapter seems to suggest that negotiating a uniform minimum price on carbon can have several desirable properties, including especially, helping to internalize the global warming externality. To fully defend the relative practicality of what I am proposing would probably require a book, not a chapter. In any event, *this* chapter is not primarily about practical considerations of international negotiations. I leave that important task mostly to others.[8] However, I do want to mention just a few real-world considerations that have been left out of my mental model, yet seem especially pertinent.

An example of a relatively small practical issue that I am waving aside is just where in the production chain a carbon price should be collected. I think the presumption would be that the carbon price should be collected by the country in which the carbon dioxide is actually released into the atmosphere. One might try to argue that a carbon price should be collected downstream as close as possible to the point where the carbon is burned. But this would involve an impractically large number of collection points. It is much easier to collect the price upstream, at various choke points where the carbon is first introduced into the carbon-burning economy.[9]

A truly critical issue is that a binding international agreement on a uniform minimum carbon tax or price requires some serious compliance mechanism. To begin with, the carbon price must be observable. For enforcement, perhaps there is no practical alternative to using the international trading system for applying tariff-based penalties on imports from noncomplying nations. Nordhaus (2015) advocates such an approach with uniform border tariffs on imports from nonmember countries imposed by a "climate club" of member nations who agree to impose on themselves a harmonized carbon price. Cooper (2010) has argued for an expansive interpretation whereby the internationally agreed charge on carbon emissions would be considered a cost of doing business, such that failure to pay the charge would be treated as a subsidy that is subject to countervailing duties under existing provisions of the World Trade Organization.[10]

An efficient carbon price naturally produces more winners than losers (by the metric of the modified Pareto criterion). In the case of the global warming externality, which has been characterized as the greatest public goods problem of all time, it seems reasonable to suppose that there might be many times more winners than losers from imposing a uniform carbon price. Because countries here get to keep their own carbon-price-generated revenues, then welfare-compensating transfers, to the extent they are made at all, ought, at least for small changes, to be relatively modest second-order deadweight-loss triangles instead of the relatively immodest first-order rectangle transfers associated with tradable permits from, say, an initial assignment of caps that are equal per capita.[11]

I close by noting again that global warming is an extremely serious as-yet-unresolved international public goods problem. With the failure of a Kyoto-style quantity-based approach, the world has seemingly given up on a comprehensive global design, settling instead in the 2015 Paris COP21 Agreement for completely voluntary and sporadic national, subnational, and regional "contributions." These partial measures seem far from constituting a socially efficient response to the global warming externality. Perhaps, as was previously suggested, a quantity-based focus on negotiating emissions caps embodies a bad design flaw. The arguments of this chapter suggest a way in which negotiating a binding internationally harmonized nationally collected minimum tax or price on carbon emissions might help to internalize the global warming externality by empowering an "I will if you will" approach.

Notes

1. One ton of carbon equals 3.67 tons of carbon dioxide. My default unit is carbon dioxide (CO_2).

2. One could try to argue that binding green fund equity payments are required to get n countries to agree in the first place to negotiate a uniform

carbon price, also representing an n-dimensional problem. However, in Weitzman (2014) I suggest that the required green fund payments may be smaller than the absolute value of the (positive or negative) transfers involved in a cap-and-trade regime that starts off, say, with equal per-capita permit assignments. Cramton, Ockenfels, and Stoft (2015) argue additionally that choosing a green fund equity–payment formula for a uniform price can itself be reduced to a one-dimensional focal problem.

3. Schelling (1960). See also the special 2006 issue of the *Journal of Economic Psychology* devoted to Schelling's psychological decision theory, especially the introduction by Colman (2006). Three of the seven articles in this issue concerned aspects of focal points, testifying to the lasting influence of the concept.

4. Coase himself did not invent or even use the term "transactions cost" but he prominently employed the concept. For an application of the transactions cost approach to controlling greenhouse gas emissions, see Libecap (2013).

5. In Weitzman (2014), I discussed negotiating one worldwide aggregate emissions cap (*contingent* upon a previous round subdivision formula for n fractional targets, set, for example, by a preceding agreement on various target reductions from various baselines). A system based on negotiating aggregate emissions (*given* a subdivision formula) could, in principle, embody a countervailing force against the global warming externality. But, again, I concluded that negotiating the extra layer of n first-round Kyoto-like fractional subdivision target reductions will likely flounder politically when applied on a worldwide scale.

6. See Weitzman (2014).

7. A minimum carbon price could theoretically be attained in a cap-and-trade system by setting it as a floor that could be enforced by making it a reserve price of permits actualized by a hypothetical international agency that buys up excess permits whenever the price falls below the floor. (Alas, such a mechanism invites its own free-rider problem, because each nation has an incentive not to spend its own money, but for *other* nations to spend *their* money to buy up excess permits.) Alternatively, a hypothetical worldwide consignment auction for carbon permits with a uniform reserve price might work in theory but seems highly impractical in practice. Again here, there is a marked distinction between the simplicity of a one-dimensional price tax and the complexity of negotiating an n-dimensional quantity-based binding agreement among n different nations.

8. See, e.g., Bodansky (2010) or Barrett (2005).

9. This set of issues and its distributional consequences (including references to other literature) is discussed extensively in Asheim (2012).

10. See also the discussion of the legality of such sanctions under the WTO provisions in Metcalf and Weisbach (2009).

11. Cramton, Ockenfels, and Stoft (2015) make an analogous argument in the form of a numerical example indicating that committing to a price tends to be less risky than quantity targets. Thus, according to this reasoning, equity transfers under cap and trade would have to be larger than equity transfers under a uniform price because of the increased risk imposed by caps. In a separate argument, they also indicate that choosing a particular green fund equity–payment formula to encourage participation in a uniform price regime can itself be reduced from a seemingly n-dimensional problem to a one-dimensional focal problem.

References

Asheim, Geir B. 2012. "A Distributional Argument for Supply-Side Climate Policies." *Environmental Resource Economics* (August 11): published online.

Barrett, Scott. 2005. *Environment and Statecraft: The Strategy of Environmental Treaty Making*. Oxford: Oxford University Press.

Bodansky, Daniel. 2010. *The Art and Craft of International Environmental Law*. Cambridge, Massachusetts: Harvard University Press.

Colman, A. M. 2006. "Thomas C. Shelling's Psychological Decision Theory: Introduction to a Special Issue." *Journal of Economic Psychology* 27 (5): 603–608.

Cooper, Richard N. 2010. "The Case for Charges on Greenhouse Gas Emissions." In *Post-Kyoto International Climate Policy: Architectures for Agreement*, edited by Joseph Aldy and Robert Stavins. New York: Cambridge University Press.

Cramton, Peter, and Steven Stoft. 2012. "Global Climate Games: How Pricing and a Green Fund Foster Cooperation." *Economics of Energy and Environmental Policy* 1 (2): 125–136.

Cramton, Peter, Axel Ockenfels, and Steven Stoft. 2015. "An International Carbon-Price Commitment Promotes Cooperation." *Economics of Energy and Environmental Policy* 4 (2): 37–50.

Goulder, Lawrence H., and Andrew R. Schein. 2013. "Carbon Taxes vs. Cap and Trade: A Critical Review." *Climate Change Economics* 4 (3): 1–28.

Libecap, Gary D. 2013. "Addressing Global Environmental Externalities: Transaction Costs Considerations." *Journal of Economic Literature* 52 (2): 424–479.

Metcalf, Gilbert E., and David Weisbach. 2009. "The Design of a Carbon Tax." *Harvard Environmental Law Review* 33 (2): 499–556.

Nordhaus, William D. 2015. "Climate Clubs: Designing a Mechanism to Overcome Free-Riding in International Climate Policy." *American Economic Review* 105 (4): 1339–1370.

Pizer, William. 1999. "Optimal Choice of Policy Instrument and Stringency Under Uncertainty: The Case of Climate Change." *Resource and Energy Economics* 21, 255–287.

Schelling, Thomas C. 1960. *The Strategy of Conflict*. Cambridge, Mass.: Harvard University Press.

Weitzman, Martin L. 2014. "Can Negotiating a Uniform Carbon Price Help to Internalize the Global Warming Externality?" *Journal of the Association of Environmental and Resource Economics* 1 (1/2): 29–49.

CHAPTER NINE

Addressing Climate Change

Does the IMF Have a Role?

MAURICE OBSTFELD

UNDER CHRISTINE LAGARDE'S leadership, the International Monetary Fund (IMF) has begun to highlight systematically the macroeconomic threats from climate change and the desirable policy responses. In November 2015, Managing Director Lagarde set out a rationale for the Fund's attention to climate change and described a range of areas in which it is engaging (IMF 2015).

The Fund's recognition of climate change as a global concern requiring global responses is by no means new. For example, the April 2008 *World Economic Outlook* devoted a chapter to analyzing the macroeconomic and financial costs of efficient policies for mitigating climate change (IMF 2008). Prior to the Copenhagen Climate Conference (COP15), Carlo Cottarelli, then head of the Fund's Fiscal Affairs Department (FAD), explained the economics of the climate externality and made the case for a "clear, credible, and broad-based carbon pricing strategy . . ." (Cottarelli 2009). This analysis has been the basis for FAD's subsequent important work on the overall costs of fossil fuel subsidies. Most recently, the October 2017 *World Economic Outlook* (IMF 2017) presented research showing that climate change will have an especially big impact on low-income countries—with the median member of this group losing nearly 10 percent of its gross domestic product (GDP) by 2020 under a business-as-usual scenario. These are the very countries that are least able to afford income losses, of course, and they have trivial levels of industrial greenhouse gas (GHG) emissions compared with richer countries. Yet, the fallout from a sharp worsening in their economic conditions will be felt the world over.

Since the Fund's earliest work on climate change, as financial crises have come and gone, the worrisome effects of anthropogenic climate change have steadily become more visible. Only in December 2015, through the historic agreement by 195 nations at the Paris climate conference (COP21), did solid hope for concerted international action emerge. Now that the Paris Agreement has entered into force (as of November 4, 2016), COP22 and future COPs face the challenge of operationalizing and strengthening the Paris undertakings over time. There remains some controversy, however—outside the Fund and even within—as to the institution's proper role in this ongoing process, one so important for the future of humankind. Here, I wish to make the case that the Fund should play a central role, not only by continuing to highlight the macro-critical aspects of climate change, but by helping countries to adapt and to mitigate in line with their commitments to the community of nations.

I see ten reasons that justify a role for the IMF in addressing climate change:

1. *Mitigating economic coordination failures.* Deep in the Fund's DNA is a desire to avoid collectively suboptimal policies. One immediate rationale for an IMF was to avoid competitive policy practices, which had worsened the Great Depression as countries pursued myopic visions of self-interest at the expense of the global community. The Nash equilibrium of the 1930s included tariffs, exchange restrictions, and competitive currency depreciation—all in the service of a mercantilist pursuit of external surpluses. The IMF's founding Articles of Agreement aimed to steer countries to a collectively superior set of policy choices. Allowing unpriced GHG emissions in pursuit of economic growth, partly at the expense of other countries, may indeed be the *mother of coordination failures,* in that the ultimate macroeconomic costs are very likely cataclysmic if business goes on as usual. The Fund can

promote global economic stability by pushing countries to recognize their impact on the global commons, and to act accordingly.

2. *The optimal solution to the coordination failure rests on economic policy.* It is well known that the lowest-cost means to address climate change is through an appropriate price on emissions, for example, taxes on carbon emissions equal to the social cost of carbon (for an early and crystal clear exposition of the general reasoning, see Solow 1971). The right price is critical not just for static resource allocation; it also provides dynamic optimal incentives for development of clean alternative energy sources. The Fund has already taken a leading role in pointing to the fiscal costs of fossil fuel *subsidies* and pushing for their phased elimination. But in its routine promotion of *smart* fiscal policies that protect public balance sheets while protecting growth, the Fund has a role in pointing out the positive spillovers and *spillbacks*— the feedback effect on oneself owing to the spillover effect—on growth from rational carbon pricing. The Fund's analysis of the short-run domestic macroeconomic impacts of such policies, alluded to earlier, can help governments evaluate the intertemporal trade-offs that carbon pricing involves.

3. *Because emissions policies are costly to implement and poorer countries will feel those costs more intensely, some resource flows to aid their efforts are appropriate to support global cooperation in reaching a more efficient allocation.* For this reason, developed countries have committed to provide developing countries with $100 billion per year by 2020 to promote emissions mitigation and climate change adaptation. And it is in their long-run interest to do so. This flow of resources, however, will impact balance of payments, investment, and production patterns.

4. *Climate change is a potent source of economic shocks.* Examples abound already. We have seen droughts with negative macroeconomic impacts from the American Pacific Rim to Morocco to Ethiopia. Extreme weather events, such as hurricanes, may now be more intense due to climate change. Ocean level rise threatens low-lying regions from Florida to Bangladesh. Ocean warming and acidification threaten the destruction of coral reefs, fish supply extinctions, waterborne disease proliferation, and frozen seabed methane release as detailed in a recent report from the International Union for Conservation of Nature (IUCN 2016).

In general, recent economic research underscores how economic productivity begins to decline as temperatures rise beyond annual averages of about 13°C (see the work by Burke, Hsiang, and Miguel 2015, extended in IMF 2017). Earlier work by Burke et al. documented how rising temperatures systematically exacerbate human conflict (Burke, Hsiang, and Miguel 2013). Naturally, all these costs fall most heavily on poorer countries, as noted above, but richer countries are far from immune. Climate change has the potential to trigger mass migrations out of stricken poor countries with potent spillovers for the rich.

5. *In line with Sustainable Development Goal targets, adaptation to climate change is critical for macroeconomic resilience.* The Fund can play a role in assessing these efforts and their macro consequences. To promote the integration of climate and energy issues into its regular Article IV surveillance, the Fund established pilot programs in nearly twenty countries ranging from Angola to the United States. While primarily concerned with issues around carbon pricing and fiscal consequences, the pilots also touched on vulnerability analysis and adaptation. The Fund also has ongoing work on small states' resilience to natural disasters and climate change.

6. *Aside from the direct effects of GHG emissions on global temperatures, emission reduction yields co-benefits for health.* These have major direct implications for welfare and labor productivity, as well as for public budgets, and certainly influence the domestically efficient carbon price. For an analysis of co-benefits from the Fund, see Parry, Veung, and Heine (2014).

7. *Low global investment has been a drag on aggregate demand.* Investments in new green technologies, as well as in adaptation, can lift demand in an environment of tepid global growth. Morocco's investments in wind parks and solar plants, which are planned to raise its renewable share of energy production to 42 percent by 2020 and 52 percent by 2030, are a case in point.

8. *For such investments, passing the cost-benefit test depends on a discount rate that reflects macroeconomic phenomena.* In general, the discount rate for risk-free investments will depend on expected future economic growth as well as stochastic elements. But climate investments pay off most strongly through avoidance of disasters, such as those related to tipping points—and the nature

of the appropriate discount rate therefore is more complex. But this again is a macroeconomic issue (as illustrated by Christian Gollier's contribution in chapter 3 of this volume), and thus is within the legitimate purview of the IMF's analytical work.

9. *Climate risks imply financial stability risks, and finance for green investment faces market obstacles.* Bank of England Governor Mark Carney famously raised the question of stranded assets in a 2015 speech—and was roundly criticized for overstepping his remit. But he was right. It is legitimate to ask if asset prices fully incorporate such risks, or if other financial contracts such as insurance contracts are appropriately priced. And the answers have implications for investment strategies, as explained, for example, by Andersson, Bolton, and Samama (2016). Jean Boissinot and Frédéric Samama discuss some of these issues in chapter 12 of this volume as well as mobilization strategies for green finance.

10. *Monitoring the nationally determined contributions (NDCs).* Only the IMF carries out annual economic health checkups—the Article IV consultations—for 189 countries. These missions would therefore provide a unique opportunity to monitor and publicize progress toward meeting the Paris mitigation commitments (the nationally determined contributions, or NDCs). This scrutiny could occur alongside other elements of climate and energy policy already incorporated in the pilot programs the Fund has successfully mounted. Such surveillance is squarely within the Fund's remit, as success in meeting and strengthening the NDCs is surely of macro-critical importance.

Even if one accepts only a subset of the preceding answers, the case for IMF involvement in addressing climate change is powerful. If one accepts most or all of them, the case is, to my mind, overwhelming.

References

Andersson, Mats, Patrick Bolton, and Frédéric Samama. 2016. "Hedging Climate Risk." *Financial Analysts Journal*, no. 72 (May/June), 13–32.

Burke, Marshall, Solomon M. Hsiang, and Edward Miguel. 2013. "Quantifying the Influence of Climate on Human Conflict." *Science* 341 (September 13): 790–794.

Burke, Marshall, Solomon M. Hsiang, and Edward Miguel. 2015. "Global Non-Linear Effect of Temperature on Economic Production." *Nature* 527 (November 12): 235–239.

Cottarelli, Carlo. 2009. "Climate Change: Some Simple (and Quite Convenient) Truths." *iMFdirect* (blog), November 23. URL: https://blog-imfdirect .imf.org/2009/11/23/climate-change%e2%80%94some-simple-and-quite -convenient-truths/.

IMF (International Monetary Fund). 2008. "Climate Change and the Global Economy." Chapter 4 in *World Economic Outlook* (April).

IMF. 2015. "The Managing Director's Statement on the Role of the Fund in Addressing Global Climate Change." November 25.

IMF. 2017. "The Effects of Weather Shocks on Economic Activity." Chapter 3 in *World Economic Outlook* (October).

IUCN (International Union for Conservation of Nature). 2016. *Explaining Ocean Warming: Causes, Scale, Effects and Consequences.* Gland, Switzerland: IUCN.

Parry, Ian, Chandara Veung, and Dirk Heine. 2014. "How Much Carbon Pricing Is in Countries' Own Interests? The Critical Role of Co-Benefits." IMF Working Paper WP/14/174 (September).

Solow, Robert M. 1971. "The Economist's Approach to Pollution and Its Control." *Science*, no. 173 (August 6), 498–503.

CHAPTER TEN

Post-Paris Clean Energy Options for China

UJJAYANT CHAKRAVORTY,
CAROLYN FISCHER, AND
MARIE-HELENE HUBERT

Introduction

The global landscape of energy emissions has evolved dramatically since 1990. At that time, North America and Europe accounted for 45 percent of global carbon emissions, while China represented 11 percent. By 2012, Chinese emissions were 28 percent of the global total, roughly equal to those of North America and Europe together.

China is home to 20 percent of the world population and one of the world's fastest growing economies. Over the past three decades, Chinese annual carbon emissions have increased from 2.5 billion tons of CO_2 to more than 9 billion tons in 2011 (World Bank, 2016) (figure 10.1). In 1990, the average Chinese citizen emitted 2 tons of CO_2 per year; three decades later, this number has gone up threefold.

Amidst this evolving landscape of energy demand, the Paris Agreement negotiated under the United Nations Framework Convention on Climate Change (UNFCCC) requires all parties to take action appropriate to their circumstances and "put forward their best efforts" through nationally determined contributions (NDCs) and to strengthen these efforts in the years ahead. Given its prominence and growth as an energy-related greenhouse gas emitter, great attention is being paid to

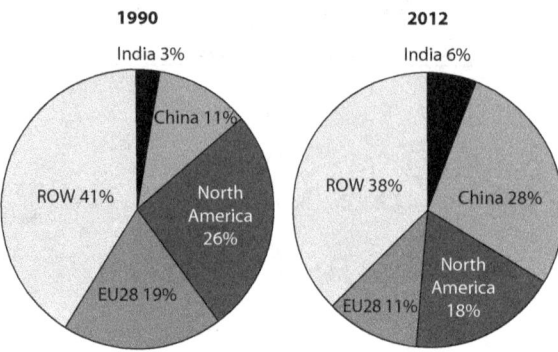

Figure 10.1 Shares of global GHG emissions. *Source*: IEA (2015).

China's NDC. The Chinese targets for 2030 include lowering the carbon intensity of their economy by 60–65 percent compared to 2005, increasing the share of non-fossil fuels to 20 percent, and peaking their emissions by 2030.[1]

To meet these targets, the main challenge for China is to reduce its reliance on coal as a primary energy source. Currently, 76 percent of Chinese electricity production is generated with coal, as compared to 46 percent worldwide. While China has comparable shares of hydro and biomass power to worldwide averages (17–18 percent) and renewables from solar and wind (3–4 percent), it lags far behind in natural gas use (2 percent, as compared to 21 percent worldwide) and nuclear (2 percent, as compared to 11 percent worldwide) for generation (figure 10.2).[2]

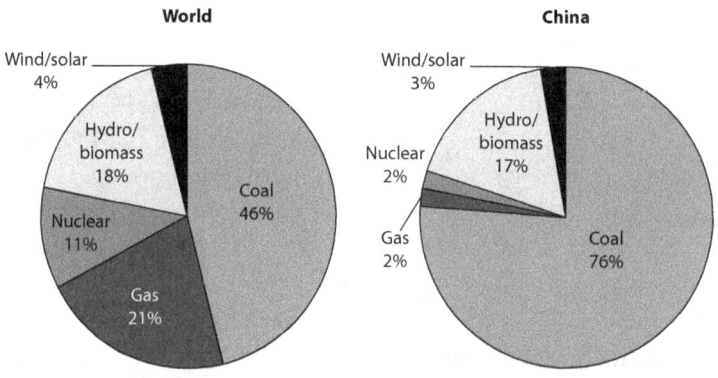

Figure 10.2 Electricity generation shares by fuel sources. *Source*: IEA (2016).

Although China's share of renewable energy in power generation is comparable to the global average, its energy growth has been such that it now has the world's largest installed capacity. China has installed 40 GW of solar photovoltaic (PV) and 145 GW of wind power, roughly double that in the United States (that has installed 18 and 74 GW, respectively) (figure 10.3). In 2015, China led the world in clean energy investment with a record $90 billion invested in renewables in

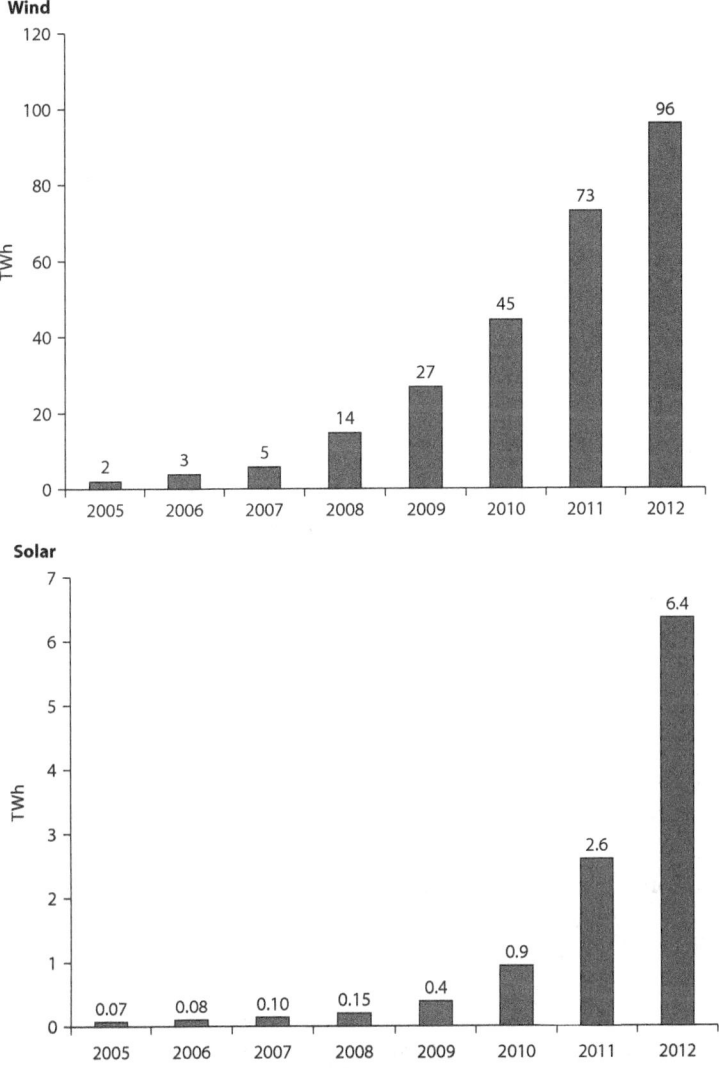

Figure 10.3 Chinese installed capacity. *Source*: EIA (2015).

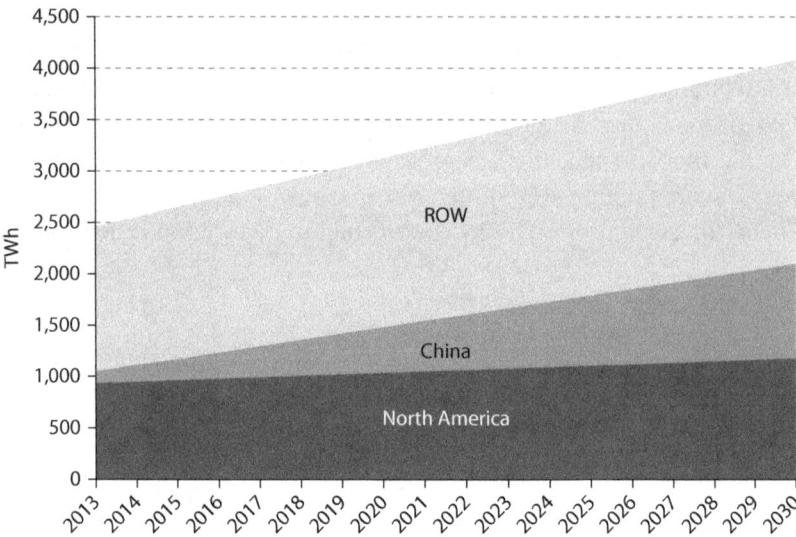

Figure 10.4 Planned nuclear capacity. *Source*: Linear extrapolation from IEA, World Energy Outlook 2014 (IEA 2014).

2014, which corresponds to a 32 percent increase compared to a 2013 level (McCrone 2015). However, in spite of the high installed capacity in renewables, regulatory and institutional inertia inhibits the full dispatch of clean power at the expense of coal and other polluting fuels.[3]

Nuclear power is also growing faster in China than anywhere in the world. More than a third of all nuclear reactors planned globally are in China, and it expects to grow its nuclear generation share to 20 percent by 2030 (or 1065 TWh) (figure 10.4).

The question is: How effective is this nuclear strategy for reducing coal reliance in China? To what extent will nuclear expansion crowd out renewable energy instead? How do learning rates in new renewable energy sources affect the substitution of nuclear for other energy sources? Is there a greater role for carbon pricing in China's energy transition and in meeting its Paris commitments?

Research Approach

A simple three-region model can help us study the Chinese clean energy transition. In recent research, we built a dynamic, partial equilibrium model representing China, North America (United States, Canada, and Mexico), and the rest of the world (ROW). We chose a partial rather than a general equilibrium model so that the relationships between the model parameters and outcomes are transparent, and in any case, we think the general equilibrium effects may be small and second order relative to changes in the energy sector of these economies. We include three fossil energy resources (coal, oil, natural gas), new renewables (solar and wind), other renewables (biomass and hydro), and nuclear power. The three main sectors of the economy consume energy: transport, industrial, and residential/commercial consumers. Final energy can be produced domestically from a combination of electric and nonelectric energy. Nonelectric energy represents the direct use of fossil fuels and renewable sources to produce final energy. Electricity is produced from a convex combination of individual fossil fuels, renewable sources, and nuclear generation. Figure 10.5 shows these relationships for each of the three economies.

Fossil fuels are traded internationally, and they have upward-sloping curves in each region, calibrated from International Energy Agency (IEA) data. Since energy prices differ across regions, transportation costs are calibrated to equal those price differentials in the baseline.

While fossil and new renewable energy sources are assumed to be determined by competitive market forces, we assume other energy sources are fixed. The combination of hydro and biomass power is fixed

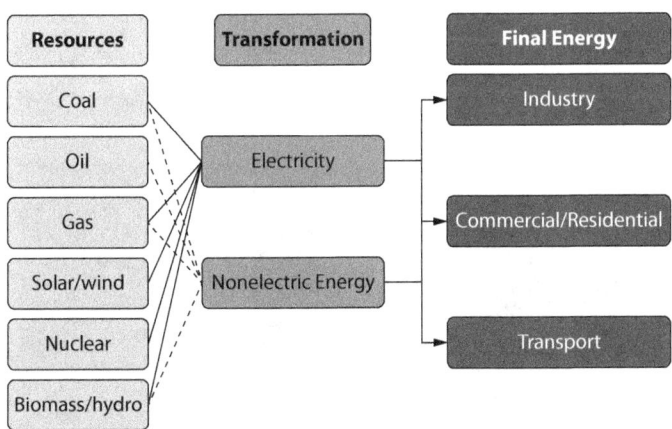

Figure 10.5 Schematic of the model.

at current levels since it is unlikely to grow significantly. It is difficult to build large-scale hydropower that will raise its share in the total energy mix significantly. Although hydropower may grow in the future, it may be a small share of the energy grid because of its remote location away from most energy demand centers.[4] Nuclear energy requires significant long-term planning; our scenarios either fix nuclear energy at IEA projections or consider a complete freeze of nuclear energy at current levels.

New renewables experience learning by doing (LBD), in which generation costs fall as a function of cumulative production. We group solar and wind energy together as new renewables. In our central scenario, a doubling of production decreases average costs by 13 percent (IEA 2014). Operation and maintenance costs are assumed to decrease by 1 percent annually.

Sectoral demand is a function of regional gross domestic product (GDP) that is assumed to have an exogenous growth path over time; and the price of energy, which is endogenous to the market equilibrium and policy variables. Figure 10.6 displays the GDP assumptions,

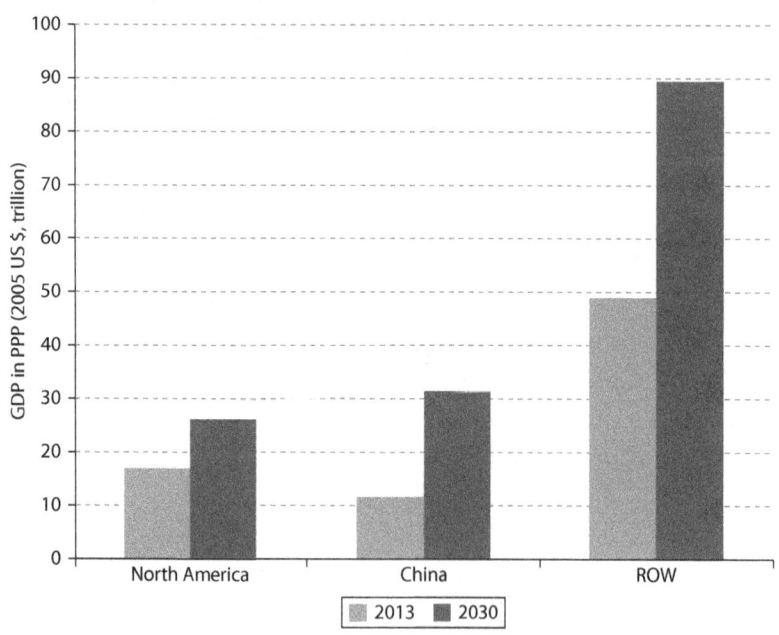

Figure 10.6 GDP in 2013 and 2030. *Source*: EIA (2015).

expressed in constant 2005 U.S. dollars with purchasing power parity (PPP), drawn from the Energy Information Administration (EIA 2015).

The model was initially calibrated for the base year 2013 and it is projected until 2030. The discount rate assumed is 2 percent.

The model works as follows. The social planner maximizes the consumer and producer surplus given resource availability and technological and market equilibrium constraints. The model parameters are calibrated to replicate as closely as possible the values of the endogenous variables in base year 2013. It then runs dynamically to predict prices and quantities for each year as well as resources used and carbon emissions generated. The use of each resource in the production of electricity and nonelectric energy are also endogenous in our model. This enables us to make projections on the future level of carbon emissions in each region. Finally, our model can project future sectoral demand as well its price. Under the baseline scenario in absence of any climate policy, the growth in demand for energy is driven by GDP growth, which is exogenous. The quantity of each resource used in the energy portfolio depends on the relative price of each resource. When energy demand increases, more expensive energy sources like unconventional resources and new renewable sources become competitive.

Alternative Energy Futures

We consider the interplay between two pairs of policy scenarios: the continued buildup of nuclear capacity in China and a carbon intensity reduction as part of the Chinese commitment in Paris.

Policy 1: Chinese Nuclear Generation Capacity Growth

In the 2014 *World Energy Outlook* which focuses on nuclear energy, the *high nuclear* case predicts that world-installed nuclear capacity is expected to grow to 767 GWe from the current installed capacity of 380 GWe. We interpolate a uniform annual growth of nuclear capacity from 2013 to 2040 of 14.9 GWe. By applying this to the nuclear capacity in the baseline year 2013, we get a nuclear capacity of 618 GWe in 2030. This increase is heavily concentrated in China (46 percent), India, Korea, and Russia (altogether 30 percent) and the United States (16 percent), along with a 10 percent drop in the European Union (EU). The annual growth in Chinese electricity output from nuclear energy is 47.65 TWh, and 14.30 TWh in North America. Finally, in the ROW, the annual growth in electricity output is 33.35 TWh. Figure 10.4 shows the projected growth in electricity output produced from nuclear energy, which is expected to reach 1029 TWh in 2030.

Policy 2: Freeze in Chinese Nuclear Capacity at 112 TWh

Following the Fukushima disaster in 2011, many countries have suspended their nuclear programs and are freezing or even decommissioning capacity. For example, Germany committed to phasing out nuclear energy by 2022; an Italian referendum overwhelmingly signaled disapproval of building nuclear capacity; Switzerland and Spain have banned new reactor construction; and Sweden will allow no more than the replacement of retired capacity. Capacity additions in the United States are stalled for a variety of reasons, and even China suspended approvals of new reactor construction pending a review of nuclear safety. Thus, we consider a *low nuclear* case, following that of the IEA, which assumes that world nuclear capacity should decline from the current level of 380 to 366 GWe. In this scenario, we assume that nuclear capacity in China is held constant at current levels, as does the supply of electricity from nuclear plants.

Policy 3: Chinese Emissions Intensity Targets

In this scenario, we model a 65 percent intensity reduction from 2005 in 2030, as delineated in China's Paris commitments. This target is imposed as a constraint on the model in year 2030. Since our model takes GDP as fixed (see figure 10.6), the constraint on CO_2 intensity can be translated into a CO_2 emissions cap by multiplying by GDP. In 2005, the carbon intensity was 1.07 kg of CO_2 per dollar; a 65 percent decline thus implies a carbon intensity target of 0.38 kg of CO_2 per dollar for 2030 (see table 10.1). Assuming a GDP of $31,430 billion as projected by the EIA, Chinese carbon emissions are restricted to no more than 11.94 billion tons of CO_2 in 2030. For reference, Chinese CO_2 emissions in 2013 equaled 9.83 billion tons (IEA 2015).

TABLE 10.1
Carbon Intensity, GDP, and Emissions in 2030 with Intensity Target

	2005	2030
Carbon intensity (kg CO_2/2005$)	1.07	0.38
GDP (2005$, billion)	5387	31,430
Emissions (billion tons of CO_2)	5.56	11.94

Sources: GDP is from EIA (2015) and carbon emissions are from IEA (2015); carbon emissions in 2030 are calculated to achieve a 65 percent decline in carbon intensity in China in 2030.

Baseline Scenario: Nuclear Growth and No CO_2 Targets

The baseline results from the model, which assume the expected growth in nuclear-installed capacity and no other restrictions (i.e., Policy 1 alone), suggests that Chinese CO_2 emissions will grow significantly, but because of a concomitant rise in GDP, the carbon intensity will decline from 2005 levels by about 49 percent (see table 10.2).

The baseline energy mix is shown in figure 10.7. There is some decline in coal use, and increase in the share of oil and renewables. Carbon intensity declines significantly but falls significantly short of China's commitment of a 65 percent reduction from 2005 levels.

Comparing Future Scenarios

Next, we consider the interplay between the different nuclear policies and the presence or absence of carbon pricing. Table 10.3 shows the carbon intensity and aggregate emissions under each of the scenarios. When we impose the 65 percent intensity reduction target on the economy, the constraint introduces a shadow price on carbon and thus makes fossil fuels costlier relative to clean fuels. Moreover, by increasing the price of

TABLE 10.2
Baseline CO_2 Emissions and Carbon Intensity

	2013	2030
Carbon intensity (kg CO_2/2005$)	0.98	0.50
Emissions (billion tons of CO_2)	9.83	15.80

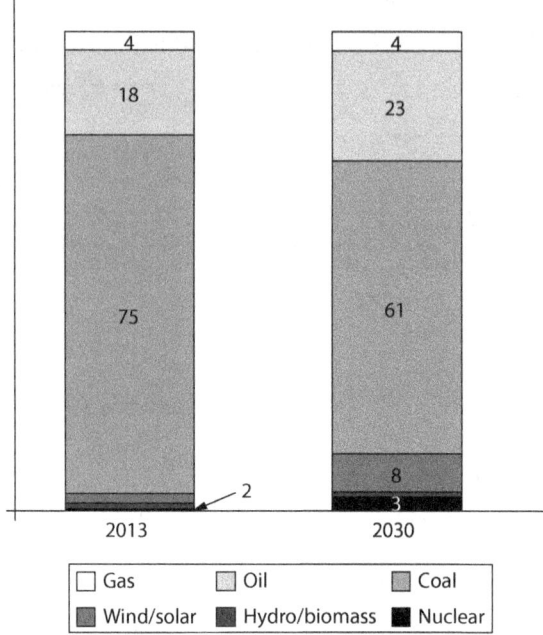

Figure 10.7 Baseline energy mix.

Note: Figures represent shares of each resource in energy consumption.

energy, demand for energy is reduced; hence carbon emissions go down significantly from over 15 billion tons to under 12 billion tons annually. The intensity target introduces an implicit tax of about $75 per ton of CO_2. This figure is robust to assumptions about the growth of nuclear

TABLE 10.3
Emissions and Intensity Under Different Scenarios

	CO_2 Emissions (billion tons of CO_2)	Carbon Intensity (kg/CO_2/$)	Carbon Tax ($/ton)
Nuclear Growth, No Intensity Target	15.80	0.50(−53%)	—
Nuclear Growth, with Target	11.94	0.38(−65%)	74
Nuclear Freeze, No Intensity Target	15.80	0.50(−53%)	—
Nuclear Freeze, with Target	11.94	0.38(−65%)	75

Note: Carbon intensity in 2005 = 1.07. Reductions relative to 2005. China's Paris Commitment: 65 percent reduction from 2005 levels.

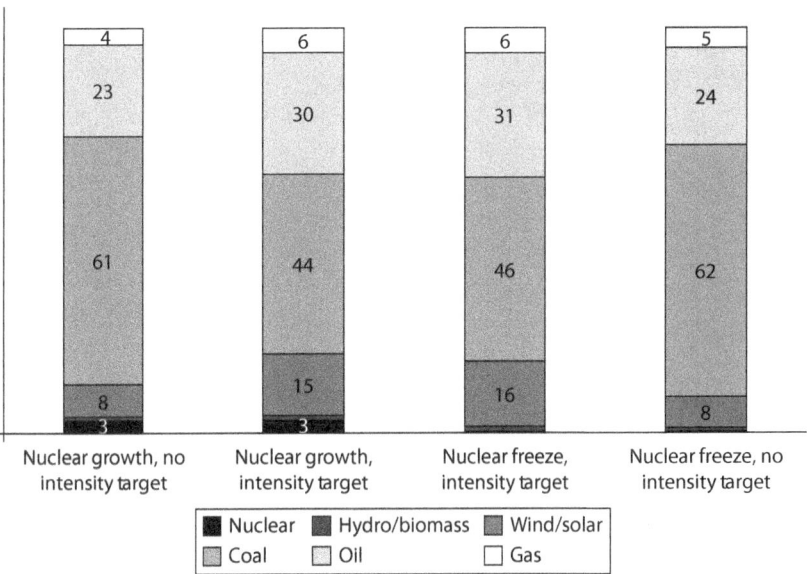

Figure 10.8 Energy mix under alternative climate goals.

Note: Figures represent shares of each resource in energy consumption.

capacity—a freeze on nuclear versus the realization of planned nuclear capacity does not change the cost of implementing the intensity target.

What is clear from figure 10.8 is that the share of renewables reaches 8 percent of the energy mix when nuclear expansion is frozen, then rises to 15–16 percent under a carbon intensity target. At least given our assumptions on the degree of learning by doing in the new renewables, there is a sharp increase in renewable use only under a carbon tax, not otherwise. The carbon price is the policy that makes renewables more competitive relative to fossil fuels, and it does so in all sectors. By contrast, nuclear energy, though clean, crowds out both renewable and nonrenewable sources of electricity generation, and it does not influence the energy mix directly in any other sector.

The Role of Renewables

Recall that with learning by doing, the cost of renewable energy is determined by the cumulative installed capacity of these renewables: the larger the magnitude, the lower is the average cost. Thus, the

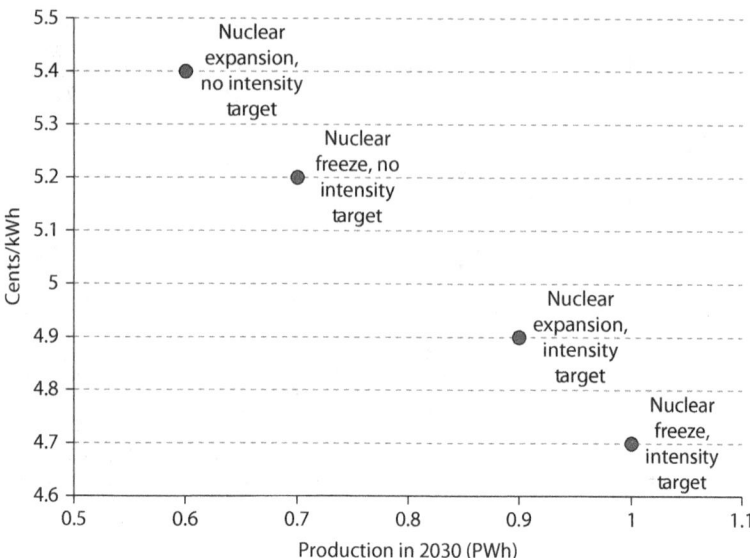

Figure 10.9 Average cost of new renewables in 2030.

future policy context has a significant influence not only on the future reliance on renewable energy, but also its cost, which further feeds back to the future capacity installations.

Figure 10.9 depicts the average cost of the new renewables in the year 2030 by policy combination. In the case of nuclear expansion and no carbon intensity target, the cost of renewables in 2030 is 5.4 cents/kWh while a moratorium on nuclear power and a carbon intensity target reduces it to 4.7 cents/kWh. These figures assume a modest rate of learning.

How does a higher learning rate in renewables influence these results? Figure 10.10 shows the learning rates projected by the IEA for low-, medium-, and high-learning scenarios.

Suppose now that we take the high-learning scenario, so that renewable energy costs decline by 20 percent with a doubling of cumulative production of renewables. The results are shown in figure 10.11. Note that higher-learning rates make a significant difference in terms of inducing a larger share of energy—the share of new renewables rises from 8 to 23 percent when we model high-learning rates, even without a carbon target. With the intensity target, quicker learning

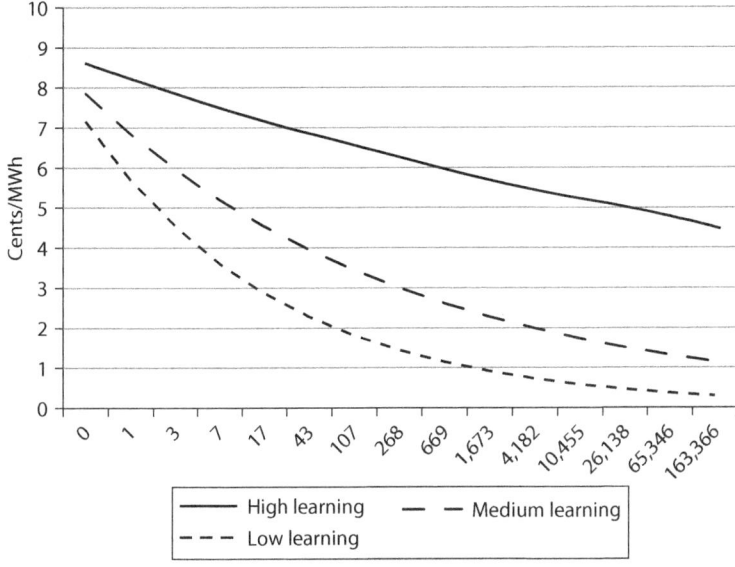

Figure 10.10 Renewable energy costs and alternative learning rates. *Source*: IEA (2014).

more than doubles the share of new renewables from 15 to 32 percent. Renewables occupy almost a third of the energy mix when high rates of learning by doing are combined with a carbon intensity target.

Increasing reliance on lower-cost renewables also influences the energy totals. With higher-learning rates, total energy consumption is higher in all scenarios than in the modest learning case. At the same time emissions are lower, as is the cost of meeting the intensity target: the implicit carbon price falls from roughly $75 to $50 a ton of CO_2 in this case.

Conclusions and Policy Implications

The success of the Paris Agreement depends critically on whether China is able to meet its stated environmental goals by the year 2030. The main goals are that the share of non-fossil fuels in the energy mix must exceed 20 percent and a reduction in carbon intensity of 65 percent from 2005 levels. As we see from our modeling exercise, the first goal can be met with some effort, since the non-fossil share

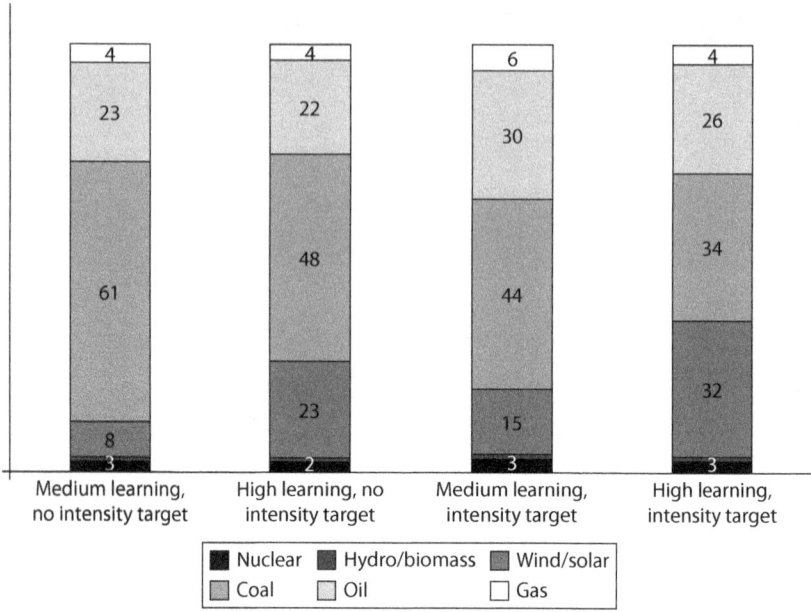

Figure 10.11 Fuel substitution in the energy sector under higher rates of learning by doing.

Note: Figures represent shares of each resource in energy consumption.

includes renewables such as hydropower and nuclear energy. Our baseline projections suggest that the business-as-usual policies lead to an 8 percent share of renewable energy in the energy mix. Only when we assume high rates of learning (20 percent for every doubling of the installed capacity) do we see the 20 percent share of renewables surpassed comfortably.

What does our analysis say about how feasible these Chinese commitments are? Well, for one, it will likely not be easy to meet the targets without putting some form of carbon pricing in place. Nuclear energy alone will not get the job done. Similarly, there are limited opportunities for large-scale building of hydropower, because most hydropower resources are located in regions such as Tibet that are not close to demand centers for energy. However, if the benefits from large cost reductions in solar and wind energy continue to be realized, one could see both targets met by the year 2030, and at reasonable prices for carbon. The massive expansion that China is currently undertaking in building new nuclear power plants is certainly useful, but it ultimately accounts for a small share of energy in the total mix.

A critical issue is the question of regulatory reform in the electricity sector in China and the role of markets more generally. Current rules for dispatch do not provide incentives to utilize lower-cost sources, which tend to favor coal-based power generation. Incentives need to be aligned so agencies that dispatch power share in the profits and favor lower-cost sources. Without such reforms, there may be a large renewable installed capacity, but it will be underutilized. There are political and institutional impediments that favor the generation of coal power even though cost savings can be achieved by switching off coal, especially at night, when wind energy is cheap. Recent reductions in renewable capacity expansions in China suggest that the government is prioritizing increased use of the built capacity rather than further additions to it.

We note several limitations to a study of this type that can be examined in future work. Probably the most important is not recognizing the role of China as a major exporter of renewable energy. If the cost reductions from the buildup of solar and wind power—indeed, the largest installed capacity in the world—translates into a large export market in solar panels, then that cumulative global experience will in turn induce further cost reductions both domestically and abroad, leading to more substitution of renewables and lower fossil fuel use. One can also consider the role of capital depreciation. Here, we assume that the renewable installed capacity does not depreciate; although that may be a reasonable assumption to make over a planning horizon of two decades or so, the evolution of the technology has been such that older wind farms and PV installations are being retired early for replacement with newer, more powerful models. Of course, fossil-based technologies also experience depreciation we have not modeled, but the typical life span of a conventional power plant is thirty to fifty years, as compared to eight to twenty for wind turbines.

Notes

1. United Nations Framework Convention on Climate Change. http://www4
   .unfccc.int/Submissions/INDC/Submission%20Pages/submissions.aspx.
2. International Energy Agency (IEA 2016).

3. See *China Dialogue* (https://www.chinadialogue.net/article/show/single /en/9119-How-China-can-stop-wasting-wind-energy).
4. See *The Diplomat* (http://thediplomat.com/2015/07/the-hidden-costs -of-chinas-shift-to-hydropower/) for issues with exploiting the large hydro-power potential for China.

References

IEA (International Energy Agency). 2014. *World Energy Outlook*. Paris: IEA.

IEA. 2015. *CO_2 Emissions from Fuel Combustion: Highlights*. Paris: IEA.

IEA. 2016. *World Energy Balance: Non OECD Countries*. Paris: IEA.

McCrone, Angus, ed. 2015. *Global Trends in Energy Investment 2015*. Bloomberg, New Energy Finance.

UNEP and BNEF (United Nations Environment Program and Bloomberg New Energy Finance). 2016. *Global Trends in Renewable Energy Investment 2016*. UNEP and BNEF Report.

U.S. EIA (U.S. Energy Information Administration). 2015. *International Energy Outlook*. Washington, DC: U.S. EIA.

The World Bank. 2016. *DataBank*. Washington, DC: The World Bank.

PART FOUR

Finance and Sustainable Infrastructure

CHAPTER ELEVEN

Financing Sustainable Infrastructure

THIERRY DÉAU AND JULIEN TOUATI

HISTORY, ECONOMIC AND social conditions shape our beliefs. For most decision makers today, the end of the twentieth-century context in which they had been educated was fairly simple in relation to infrastructure: developed markets, mainly Western Europe, Japan, and North America, had been equipped post–World War II (WWII) and the developing world needed, to extent possible, to catch up mainly thanks to the development aid from the North.

For Organization for Economic Co-operation and Development (OECD) institutional investors considering infrastructure investment in the 1990s, the situation was even simpler: as investing outside of the OECD, say in Africa, was quasi-impossible to contemplate, save in extremely limited proportions, infrastructure investment was mainly about purchasing existing assets only requiring limited maintenance efforts or utilities necessitating primarily efficiency gains. This vision of infrastructure was by far too narrow. With perhaps the exception of Japan, developed countries had actually at the time underinvested in their infrastructure.

New long-term trends, such as aging populations or environmental awareness, were in addition fueling a significant demand for greenfield, infrastructure. In the developing world, China demonstrated a capacity to attract global and local savings alongside a strong budgetary capacity to finance its infrastructure with limited reliance on traditional aid instruments (whose size and rationale was, in any case, not adapted to the investment-savvy Chinese catchup).

The fight against climate change and more widely the so-called low-carbon transition has been one of these trends disrupting this contingent consensus we have just described. This low-carbon transition is actually essentially about replacing flows of carbon (burning oil in old and relatively cheap power plants) by flows of capital (replacing these old, polluting power plants by capital-intensive renewables or investing on the demand side to reduce the consumption of energy). The greening of the electric system in Europe, for example, has actually caused a massive creation of new assets. In terms of installed capacity, OECD Europe added almost 50 GW of new capacity in 2011, the vast majority of it being renewables, versus only 10 GW in average per year in the post–World War II catchup period.

The challenge is therefore to channel capital, i.e., savings, that are primarily collected by institutional investors (life insurers and pension funds), to the energy transition. We will see how difficult it may be in the medium term in a context of technological disruption. More fundamentally, we will show that investments in renewables, however absolutely necessary, will not be enough and that our industry, infrastructure investment, will probably need to reinvent itself in the coming years, both in the developed and in the developing world, to address the sustainable infrastructure financing challenge. We will also raise the *How* question, i.e., think beyond the need analysis, the piling up of new initiatives, and the call for more investment to focus on the conditions that need to be gathered in a given environment to secure much-needed investments in low-carbon infrastructure projects.

Finally, we will provide, from our operational experience as infrastructure managers, another angle to the Lucas Paradox to identify one route that could enable savers from both the developed and developing worlds to finance low-carbon assets in the South without putting too much pressure on public sector budgets.

1. We Are Entering an Era of Significant Long-Term Investment Needs for Sustainable Infrastructure in a Context of Increasing Uncertainty

The global infrastructure financing gap is now widely acknowledged: $60 trillion must be invested in infrastructure to maintain gross domestic product (GDP) growth through 2030, according to the McKinsey Global Institute.[1] This figure is consistent with the World Bank's estimates for the next twenty years.

This is particularly true for the energy sector. According to the International Energy Agency (IEA),[2] the energy supply sector has already increased its annual investments from $600 billion in 2000 to $1800 billion in 2013. Interestingly, renewables, quasi-inexistent in 2000, accounted for only half of that increase at a slightly lower contribution than investment in the oil, coal, and gas supply industries (since then hit by the dramatic fall in commodity prices). This is only a start: low-carbon annual investments in the energy sector, according to the IEA, will have to increase from less than $500 billion to $2000 billion (in real terms), roughly the annual GDP of Italy, to be able to limit the CO_2 concentration in the atmosphere at 450 ppm. This is a significant part of the $3.3 trillion per year above the mentioned McKinsey estimate.

Given the long life span of most infrastructure assets—from fifteen to more than one-hundred years—a higher share of global savings will have to be allocated to infrastructure in coming years. The fast-growing savings managed by institutional investors—estimated by the Boston Consulting Group at $71.5 trillion in 2015—must play a central role. It is important to consider the nature of these low-carbon investments: with a few exceptions, these are all long-term investments. Payback may require more than ten years or in some cases (hydropower plants, for example) decades and operating costs are in general limited compared to initial investment (as there is no fossil fuel to pay for). The cost of capital must therefore be kept under control as, it could otherwise raise significant affordability issues for consumers or taxpayers bearing the eventual cost of these low-carbon assets. In other words, the sustainable infrastructure sector will need to attract *patient capital*: long term and relatively affordable.

The bright side is that these jumbo amounts are available. Institutional investors are managing assets close to $75,000 billion that are looking for safe havens offering reasonable returns in a context of still fairly accommodative monetary policies across the developed world. Most of these investors have long-term liabilities that enable them to consider long term, relatively illiquid, investments.

However, attracting such investors requires clear conditions to be met: predictability, stability, and strong confidence in capital preservation. As we will discuss later, this can be achieved in the infrastructure investment space. At this stage, we should specifically focus on trends in the low-carbon space that could make it difficult to fulfill these conditions of attractiveness for investors in the short to medium term.

The first trend is technological. The global low-carbon transition is happening in a disruptive technological environment. According to the IEA, costs of solar PV-utility scale plants have been divided by more than five between 2008 and 2016, and the same for batteries (some specialized engineering firms even claiming that the actual cost reduction for the latter may be actually tenfold), and in the lighting industry LEDs have seen their costs divided by more than twenty. In the meantime, nuclear energy has, in the aftermath of Fukushima in particular, faced an increase in its costs driven by enhanced safety requirements. In the future, solar PV and storage may continue their dramatic cost reduction while the cost of other technologies such as onshore wind may stabilize or decrease at a slower pace.

Depending on the actual pace of cost reduction for each technology, and as some technologies may become less competitive and even obsolete—bearing in mind that cost is not the only driver for the choice of a given technology—it will therefore be possible to find *green but stranded* assets in the medium term. This is not in line with the expectations of long-term and patient investors.

The second trend is political. Political support for low-carbon energy has proven critical and successful in recent years, in spite of some *local* crises such as the renewables burst in Spain at the beginning of this decade. This support takes in practice the form of long-term contractual arrangements (such as feed in tariffs, long-term power purchase agreements, or public-private partnership [PPP] contracts) that can also protect investors, up to a certain point, against the technological disruptions described above.[3] Long-term tax incentives (carbon tax or any form of carbon-pricing mechanisms) are entering into this category but with mixed success thus far in terms of trust building to attract long-term investors when compared to robust and long-term contracts.

It is now commonplace to say that we are entering an era of political uncertainty, marked by radical changes in political orientations, including with respect to the environment as demonstrated with the U.S. decision to withdraw from the Paris Agreement. How will the political support to the low-carbon transition evolve in the near to medium term continues to be a critical question even if signs of optimism remain.

New Challenges Imply a Shift in the Infrastructure Investment Industry

The energy infrastructure investment industry, both public (government) and private sectors (operators and private investors), has lived in a world of large-scale transactions. Equipping an economy to accommodate its energy needs required investments in *big* assets. Coal-fired power[4] plants can easily exceed one billion in investment size. This has a lot of advantages for the financial and industrial communities, in the particular the project finance community, as it attempts to *dilute* transaction costs (legal, engineering, and overhead costs, for instance) in large capital expenditures.

A first evolution by this industry was to accommodate much smaller projects in the renewable space. Projects of a few tens of millions can be financed using traditional project finance techniques, in part because they are easily standardized. For example, the utility scale solar PV market in the developed world is in the process of being commoditized from an investor point of view. This is actually only partially correct as regulatory environment (including support schemes) may evolve but the truth is that these assets are fairly predictable from a technical point of view. This is clearly not the case in most emerging markets (where permitting or public support processes still remain complex) despite recent attempts such as the International Finance Corporation (IFC) Scaling Solar programs to facilitate procurement and increase competition.

A revolution may however still be needed. The next wave of investments may focus more on energy efficiency projects, which are even smaller (a large energy efficiency project typically costs a few millions).

According to the International Energy Agency (IEA), they will represent half of the investments required by 2040 to reach the COP21 target, i.e., almost three times the investments in renewables.

Channeling long-term savings to these projects whatever the region will be critical. This means inventing schemes that probably lie somewhere between infrastructure and the consumer goods industry while enabling tailor-made solutions (as, save maybe with LEDs or appliances, most energy efficiency projects are singular). A microgrid of an ecodistrict in the developed world necessitates only $10 million of investment; in a remote rural village in Africa only a few hundreds of thousands; and in the electromobility space, equipping a large condominium with the adequate electricity distribution infrastructure for daily charge only tens of thousands.

Such amounts are barely significant for institutional investors, while taken together they may represent markets worth tens of billions of dollars of investments. In addition to the pure energy industry, this may fundamentally reshape the building industry as well as the mobility sector. Some interesting new models are emerging, such as joint ventures between industrial and financial investors targeting a myriad of small, complex projects but with a high potential of *replicability* in order to achieve a sufficient scale.

Fundamentals will in any case remain. Governments still need in particular to have a real strategic procurement strategy when it comes to infrastructure. The exercise of the intended nationally determined contributions (INDCs) in preparation of the COP 21 was a very good step toward governments taking control of their long-term strategic thinking. After Marrakech COP22, it is now time for their operational declination into actual investment programs.

More broadly, administrative capacity may have been considered as a given or there has been the belief that the private sector could entirely replace a failed public administration. This is not the case and the first good infrastructure investment for a public authority is often in human capital. Having the right project director to deliver a given project and/or policy maker to design a program strategy and support the political decision-making level within a government has always been essential for a country to be successful.

This is where the concept of PPP may require further considerations. As discussed in Arezki et al.,[5] this concept is multifaceted and several conditions are required to make an infrastructure project developed under a PPP scheme a success. In most cases, as illustrated in the Meridiam experience globally and in emerging markets (see the box), this success will often come when a "functioning triangle" between governments, development institutions, and the private sector are in place.

Illustration of the Triangle Between Governments, Development Institutions, and the Private Sector: The Meridiam Case

Meridiam was founded in 2005 as a long-term developer and investor in greenfield infrastructure with an initial focus on Europe and North America that has been since extended to Africa, Latin America, and Eurasia. Its twenty-five-year approach of infrastructure investment was backed by international institutional investors as well as public development institutions such the European Investment Bank (EIB).

From the EIB's point of view, this support was *inter alia* contributing to the achievement of the Trans-European Networks policy, as Meridiam targeted most of these projects when developed under PPP schemes.

The EIB also played a critical role, alongside commercial lenders, as a long-term lender, involved across Europe in more than fifteen Meridiam-sponsored PPP projects as well as a provider of technical assistance to public authorities across Europe to implement balanced PPP projects, a role that was boosted in the context of the Juncker Plan that created the European Investment Advisory Hub as a joint initiative between the European Commission and the EIB. In parallel, Meridiam initiated several strategic partnerships with development institutions, including the Development Bank of Japan, Agence Française de Développement and Proparco from France, and several other bilateral and multilateral development finance institutions such the European Bank for Reconstruction and Development (EBRD).

This Meridiam partnership with the EBRD has been particularly fruitful in terms of developmental impact. This partnership was a critical catalyst for development of the Turkish social infrastructure PPP market by developing and financing the first pilot project, the Adana Hospital PPP in 2015. This strategic partnership culminated in Turkey with the recent closing of the Elazig Integrated Health Campus PPP through the issuance of the first Green and Social project bond in Turkey with the aim of developing access for institutional investors willing to lend to projects. The Elazig Integrated Health Campus PPP reached financial close in December 2016.

The €360 million-worth Elazig project comprises the design, building, finance, and maintenance of an integrated health campus with 1038 beds. The campus is comprised of general practice, women and children, physical medicine, and psychiatric hospitals local in Elazig (Eastern Turkey), for an operation period of twenty-five years following a construction period of three years.

As mentioned, this project benefited from a broad public sector and multilateral support as part of a flagship program by the Turkish government to develop thirty-eight new healthcare campuses on a nationwide basis. It has strong social rationale for the project in a region with limited access to high-quality healthcare services. It included a strong focus on environment, social, and governance (ESG) aspects and more global sustainability throughout development. This formed the basis for the project bond's certification by the specialized rating agency Vigeo as "Green and Social."

Meridiam, EBRD, and MIGA (World Bank Group) jointly structured a unique credit-enhancement structure, enabling Moody's to assign a Baa2 rating (two notches above the rating of Turkey at the time). The structure entailed a combination of political and contractual breach cover from MIGA with construction and operation liquidity facilities provided by EBRD in order for the bond to remain current under extreme circumstances. Elazig is Turkey's first Green and Social project bond, aligning with the COP21 global commitment to support emerging countries' Sustainable Development Goals (SDGs). IFC (World Bank Group) and Proparco supported the bond as anchor investors, further mobilizing investors from Europe, Japan, and China.

The ambition is to replicate this experience of cooperation between the government, leading development institutions and private investors such as Meridiam in many other jurisdictions. In Africa, the same triangular philosophy is also at the core of the cooperation between the Senegalese government, its national utility Senelec, and Proparco, the French development finance institution, to develop and finance the first utility-scale solar PV projects in the country. These two 30 MW projects, Senergy and Ten Merina, reached financial close through more traditional forms of long-term loans from development finance institutions, and started construction just after the COP21 (the Senergy project having been inaugurated in June 2017).

Most development institutions are aware of this gap and are trying to support *project preparation* but actually more than funding consultancy support, the challenge is educational, i.e., to educate, coach, and train the future infrastructure professionals who will be able to integrate the complexities of these projects to make them happen. A global program on sustainable infrastructure delivery spearheaded by a few leading academic institutions located across the five continents aiming at training these public-sector professionals could be a bigger leap over the next ten years.

2. Revisiting the Lucas Paradox: In Defense of an Enhanced Green MIGA or Green Risk Mitigation Instrument

As explained in Lucas's seminal paper,[6] capital does not flow from developed countries to developing countries. The reality is the opposite. This macrofinancial reality has been verified many times on the ground. Raising capital to finance the low-carbon transition in Europe is relatively easy. It remains extremely difficult in Africa. How many African investors have we met willing to invest in the OECD only but not in their region? We have come across a well-managed African sovereign wealth fund fully invested in . . . listed OECD assets.

This reluctance has many roots but probably not a pure risk analysis rationale. We have observed assets offering 10 additional percentage points of return on equity—post-political risk insurance coverage costs—for a PV project compared to the same project in the North. Africa is, according to Moody's, the continent where the average default rate on project finance transactions is, by far, the lowest. In spite of this, global investors still prefer in general to pour capital in the same projects, located in a few OECD countries globally. Financial regulations can be blamed for encouraging bubbles. Actually, being an OECD country opens huge pools of capital as most regulations assimilate almost systematically OECD countries to *low*-risk countries.[7]

This is particularly problematic when considering a global challenge such as the fight against climate change. Climate preservation is a global public good. Avoiding a ton of CO_2 emissions in Burkina Faso should have the same collective value as in Texas. But the counterparty risk is considered as being so high that investors would prefer to invest in Texas even if the premium for investing in Burkina is

10 percentage points and well above the extra quantifiable risk for investing in such a country.

To a large extent, an investor willing to invest in a project located in Burkina contributing to the fight against climate change should face a counterparty/country risk identical to that of a Texan project in order to solve this perception gap. If this can be achieved through a risk mitigation instrument, it would potentially close the gap between the Paris goal to attract $100 billion a year from the North to the South and the reality of the budgetary constraints of the developed world.

Assuming that good projects exist in Burkina, how can this be achieved? By creating a global risk mitigation product, easy to mobilize and understood by investors, focusing on counterparty risk. This product could build on existing instruments that are already actionable. This is in particular the case with MIGA, an entity of the World Bank group, that offers long-term political risk insurance products that enable to significantly mitigate the risk of loss of capital should a government decide not to honor a financial obligation in relation to a given contract. OPIC (Overseas Private Investment Corporation), the U.S. development finance institution (DFI), offers similar instruments. Other DFIs are also able to provide enhancement facilities such as subordinated liquidity facilities—to cope with the potential payment delays of governments or public off-takers, or first-loss piece facilities.

This works particularly well for an infrastructure project structured under a PPP-type arrangement (which could be for instance the case of a wind farm receiving a tariff from a national utility over the long term). Today, MIGA supplies $2.8 billion of such political risk insurance *across all sectors.* This is significant but by far lower than what is needed for the low-carbon transition in emerging markets. And the *Enhanced Green MIGA* product needs to blend both the political risk insurance with highly rated (e.g., above AA) liquidity instruments to really meet the objective of fully aligning counterparty risk perception in the South and the North and finally be positively rated by the main rating agencies. A first example is the EBRD/MIGA enhancement product offered to the Elazig green and social bond for a Turkish hospital PPP where EBRD complemented the MIGA guarantee (see the box for more details).

Another successful climate change–driven product (the MCPP Infrastructure) designed by IFC and Allianz has allowed the latter to mobilize significant capital to co-invest alongside IFC across its portfolio while benefiting from a first-loss protection of 10 percent provided by IFC and Sida (the Swedish development cooperation agency).

It could be envisaged to adjust financial regulations that apply to institutional investors (e.g., Solvency II in Europe) to treat investments in the South benefiting from these climate-driven enhancements similarly as *safe* investments. It would imply from a regulatory standpoint that climate investments in developing markets would require much lower capital requirements than today[8] or, in most cases, become a permitted investment (lifting the restriction to invest only in OECD markets). This of course will have to be based on sufficient data and track record on performance of infrastructure projects financed under such schemes to justify such evolution of the regulation.

———

Channeling savings into productive investments, including greenfield sustainable infrastructure, will be essential for the fight against climate change. There is a historic opportunity for institutional investors and governments at the global level to secure both long-term financial stability and performance and contribute to sustainable growth fueled by efficiently managed low-carbon infrastructure.

We firmly believe that pushing the limit of *blended finance*, with an Enhanced Risk Mitigation Instrument or Enhanced Green MIGA combined with limited adjustments to financial regulations would help bridge the low-carbon investment gap in the developing world.

Notes

1. Dobbs, Richard, Herbert Pohl, Diaan-Yi Lin, Jan Mischke, Nicklas Garemo, Jimmy Hexter, Stefan Matzinger, Robert Palter, and Rushad Nanavatty. 2013. "Infrastructure Productivity: How to Save $1 Trillion a Year." McKinsey Global Institute (January).
2. IEA (International Energy Agency). 2016. *World Energy Investment*. Paris: IEA.
3. To take one example, Germany has accepted to fund since the beginning of the twenty-first century over the long term, expensive photovoltaic (PV) power plants that today would be totally out of money compared to new ones.

4. The International Finance Corporation (IFC) of the World Bank Group (WBG) has launched a program to support governments in the procurement of large-scale solar programs.

5. Arezki, Rabah, Patrick Bolton, Sanjay Peters, Frederic Samama, and Joseph Stiglitz. 2017. From Global Savings Glut to Financing Infrastructure: The Advent of Investment Platforms." International Monetary Fund (IMF) Working Paper. IMF (April), 221–261. Printed in Great Britain.

6. Lucas, Robert. 1990. "Why Doesn't Capital Flow from Rich to Poor Countries?" *American Economic Review* 80 (2): 92–96.

7. As a recent example, the Solvency II European framework for insurers evolved to introduce a specific favorable treatment for infrastructure investments but only for assets located in the Organization for Economic Co-operation and Development (OECD).

8. As an example, in the Solvency II framework, an investment in unlisted infrastructure in Africa benefiting from this more favorable treatment would see its corresponding requirement for the insurer reduced by more than 30 percent.

CHAPTER TWELVE

Climate Change: A Policy-Making Case Study of Capital Markets' Mobilization for Public Good

JEAN BOISSINOT AND FRÉDÉRIC SAMAMA

CLIMATE CHANGE HAS been looming in the background for a number of years but it has only recently come fully into the spotlight. Increasingly visible effects on the environment and growing media attention have made of climate change a pressing issue and have changed the backdrop of global summits on climate (UNFCCC, COPs, etc.). In this context, the financial sector has received a wakeup call that it may have a key role to play. With trillions needed to finance a low-carbon transition consistent with the "well below 2°C" objective of the Paris Agreement, the mobilization of financial institutions as well as regulators and supervisors could prove instrumental in this race against the clock.

In this chapter, we present the role of finance in the context of the low-carbon transition and the different critical developments that could reinforce its decisive contribution. We argue that a better understanding of climate change implications among financial institutions as well as financial innovations tackling scalability, time horizon and complexity issues, and cooperation within the financial sector have been instrumental in mainstreaming green finance. On the other hand, we highlight the potential for governments to accelerate and encourage these transformations to deliver their full potential by taking a more proactive role. By mobilizing capital markets (with the need to apply market rules like a risk management approach), policy makers may also achieve other public goods–related objectives. Thus, the untapped resource of leveraging the financial system may be critical for tacking climate change.

1. A Global Shift to Contain Climate Change

With sixteen of the seventeen warmest years on record occurring since 2001,[1] climate change is accelerating and becoming a fact of life for an increasing number of people. As scientific observation is accumulating evidence about the physical impact of climate change (from increasing global temperatures, rising sea levels, ice melting in the Arctic, the warming and acidification of the oceans, and increasing climate change–related adverse developments),[2] more and more people are experiencing its consequences in both developed and emerging economies.[3]

The fight against climate change has gained momentum within society. A sustainable way of life is an increasing aspiration, especially among millennials.[4] This is part of a greater movement and echoes a diversity of global voices (including Pope Francis,[5] Governor of the Bank of England Mark Carney,[6] and Chinese President Xi Jinping[7]) expressing both the urgency of tackling climate change and, more fundamentally, that a sound environment is a prerequisite for a thriving humanity.

Over the past few years and against this backdrop, climate change has shifted from a distant issue that should be dealt with by a handful of large carbon emitters (as in Copenhagen 2009) toward a collective responsibility. Rather than a problem to be solved through a top-down approach, climate change action (both mitigation and adaptation) is increasingly being identified as a relevant concern not only for states but also for nonstate actors (local governments and other public institutions, businesses and firms, financial institutions, communities, etc.).

A shift in the economics of renewable energy is also driving the fight against climate change. Renewable technologies are now widely available and their relative cost compared to that of fossil fuel–based new electricity generation capacity has dramatically decreased since 2010, a trend that is expected to accelerate (see IEA 2015, 2016). At the same time, progress in understanding the economics of the transition has also been pivotal: it is now well understood that the reorientation of investment implied by the low-carbon transition may prove beneficial from a growth perspective and that capital reallocation is central in this respect (Calderon and Stern 2014, 2015, 2016).

These notable paradigm shifts in experience, attitudes, perspectives, and economics were vital in achieving the COP21's Paris Agreement in December 2015. In a landmark climate change agreement, 195 parties committed to contain global warming to "well below 2°C" above preindustrial levels and took the pivotal step of building a global governance framework to address these issues. Each signatory has committed to prepare, communicate, and maintain successive nationally determined contribution (NDC) plans in which they indicate, among other things, their CO_2-reduction targets (to be later translated into low-carbon transition roadmaps). This collaborative, bottom-up approach was a real revolution compared to the previous COPs and an essential driver of success, as put forward by Giraud, Lancesseur, and Roulleau (2016). Since December 2015, there has been continuous mobilization and renewed commitment from most parties as the Paris Agreement effectively entered into force on November 4, 2016.[8]

While Nordhaus (2016) provides a sober reminder that success still requires substantive and determined action, even more so with the U.S. administration's decision to leave the Paris Agreement, the collective momentum and increasingly pressing consequences of climate change and the economics of the transition[9] suggest that acceleration can be expected.

2. Finance as a Driver for Change

Albeit arguably more quietly, a shift has also been taking place over the past few years in the financial sector. Until recently, financial institutions did not pay much attention to climate change. Environmentally aware financial products were limited to a niche market and only a very limited number of institutional investors were seriously considering climate change–related risks within their broader-risk management frameworks.

However, over the past five to ten years, a number of initiatives have started to emerge, now coalescing into a sector-wide mobilization. This has enabled policy initiatives which, in turn, have further strengthened the momentum behind the adoption of climate change as a strategic business issue for financial institutions.

Two examples of this realization are the development of the green bond market and the growing profile of climate change among asset owners.

Created in 2007 by multilateral development banks (the World Bank and the European Investment Bank), green bonds have enjoyed robust growth since then. Although the green bond market still represents a minor part of the overall bond market, it has become a fully fledged market with a full range of issuers (supranational, local governments and agencies, corporations, financial institutions, and more recently, sovereign[10] and even emerging markets[11]), maturities, and sizes as well as currencies (see Kaminker 2016 and CBI 2016 for a discussion of the market). This development has been facilitated by market participants' efforts toward developing standards notably through the development of the Green Bonds Principles in 2014 and their regular updates (see International Capital Market Association [ICMA] 2017).

Green bonds are traditional fixed-income instruments with the particularity of financing exclusively "green" projects. By investing in green bonds, fixed-income investors are "signaling" their interest in financing companies that put the proceeds of their debt issuance toward green projects (bearing the same risk exposure as a standard bond). In this respect, green bonds appear to be an innovative engagement device in the fixed-income market. This could even offer new possibilities for issuers to signal their strategic commitment to the low-carbon transition.[12]

Investors, and in particular asset owners, increasingly recognize the potential of green bonds as a means for financing the energy transition. In December 2016, asset managers, asset owners, and funds representing a combined $11.2 trillion of assets signed the Paris Green Bonds Statement, thereby encouraging the development of the green bond market to address climate change.[13]

Another significant development has been the momentum in the asset owner community toward addressing climate change. A $100 trillion engagement movement initiated by CDP (formerly the Carbon Disclosure Project) that pushed corporations to disclose their carbon impact developed into more substantial engagements from investors looking to know more about corporate action plans related to climate change. In 2015, a coalition of investors named "Aiming for A"[14] introduced the first shareholder resolutions at BP and Royal Dutch Shell's Assembly General Meetings. In these resolutions, 150 shareholders called on the companies to not only implement stress tests to assess if their business models were compatible with a 2°C target,[15] but also to develop a strategy for favoring the transition to a low-carbon economy, e.g., through investment in renewable energy. Similarly, during One Planet Summit in Paris, 225 investors representing more than $26.3 trillion under management committed to engage with the world's largest corporate greenhouse gas emitters to ensure that they are minimizing and disclosing the risks while maximizing the opportunities presented by climate change and climate policy.[16]

Additionally, commitment to the Montreal Carbon Pledge[17] has led a growing number of institutions (120 investors now representing $10 trillion of assets under management) to take a step further and add climate change–related risk analysis to their agendas: the commitment to measure and publicly disclose their carbon footprint.

Finally, the Portfolio Decarbonization Coalition, cofounded by CDP, UNEP FI, Amundi, and Swedish Pension Fund AP4 named by the United Nations in September 2014 as one of the leaders of climate action in the financial and business communities (Ban 2014), has set the example by providing a platform intended to demonstrate the feasibility, diversity, and scalability of available solutions for combating the risks associated with climate change and bringing this knowledge to the broader investor community. These shifts in the financial sector are having a significant impact. For example:

- Having gathered strong support, BP and Royal Dutch Shell's resolutions were successfully passed with over 98 percent favorable votes in both cases. By the end of 2015, BP and Shell were among the first companies to respond to the shareholder resolution by disclosing their climate action plans.[18]
- A portfolio of $800 billion (from twenty-nine investors representing a total $3.6 trillion in assets under management) is committed to being aligned with a low-carbon economy (through the Portfolio Decarbonization Coalition)[19] while the transition risks (see infra) of over $10 trillion of assets are beginning to be scrutinized (through the Montreal Pledge).

At a more fundamental level, financial sector mobilization is also enhancing the effectiveness of climate policies by better aligning capital allocation with climate change mitigation and adaptation goals (see Boissinot, Huber, and Lame 2015). As such, while largely ineffective in isolation, this mobilization is a key complement to climate action by policy makers and corporations.

The following sections examine key lessons from these developments toward leveraging on finance in the pursuit of a greater objective: the importance of framing climate change using a financial sector perspective (e.g., as part of the usual risk management framework, section 3), the importance of financial innovation (section 4), the importance of forming coalitions to enable knowledge building, sharing, and dissemination (section 5), and the role of governments as catalysts in these processes (section 6).

3. Framing Climate Change as a Financial Issue

One of the key developments of recent years is that climate change is now being perceived as part of the traditional risk management approach. As illustrated by Hong, Li, and Xu (2016), there is a growing consensus that short-term-oriented markets do not price climate change–related risks correctly; leaving long-term-oriented investors to face a market failure regarding climate change–related developments, as climate change effectively becomes a mispriced/mismanaged risk (Wolf 2014). Thus, as long-term investors cannot count on market signals, they increasingly recognize the need for a dedicated assessment of climate change–related risks. These fall into two distinct categories (TCFD 2016): physical risks[20] and transition risks.[21]

In practice, investors have started to focus on transition risks.[22] First, they consider that polluting companies are likely to be impacted by reactions to climate change in the long run. Changes in policies (from the removal of subsidies, see IMF 2013, 2015, to more stringent regulation and higher prices on carbon), and shifts in consumer or investor behavior are deemed likely to have a direct impact on some firms or industries, possibly rippling through the economy. A second strong intuition stems from the recognition that the risks associated with fossil fuel companies and their consistency with a low-carbon economy may not yet be fully integrated and priced in, opening up the possibility for market adjustments. Some analysts (e.g., CTI 2011, 2015; Caldecott, Kruitwagen, and MacDonald-Korth 2016) estimate that overall reserves play a significant role in oil and gas company valuations even though the full exploitation of these reserves would far exceed the planet's "carbon budget." This is likely to lead to "stranded assets" or even the burst of a "carbon bubble" if, after delayed action, policies are suddenly implemented to achieve the Paris Agreement objectives.

While more research is needed to better understand these climate change–related risks, another critical ingredient is the availability of relevant data to inform risk assessment. In order to foster progress in this area, the Task Force on Climate-Related Financial Disclosure (TCFD) was established by the Financial Stability Board (FSB) in December 2015 during the COP21 in response to a G20 mandate "to review how the financial sector can take account of climate–related issues."[23] Given that "one of the most significant, and perhaps most misunderstood, risks that organizations face today relates to climate change" (TCFD 2016), the Task Force seeks to help investors, lenders, and insurance underwriters identify and appropriately assess and price climate change–related risks and opportunities. To this end, it promotes better climate reporting by corporate and financial institutions and has developed voluntary and consistent climate change–related financial disclosures that can assist investors in understanding material risks. TCFD submitted its recommendations (TCFD 2017) to the FSB and G20 in July 2017.

4. Getting Smarter: Financial Innovations to Help Solve Time Horizon, Complexity, and Scalability Problems

Financial institutions are facing a series of challenges that could hinder action on climate change–related risks. First, many investors are facing a "tragedy of the horizon" (Carney 2015). This is the belief that the effects of climate change will be felt beyond their investment horizons. Indeed, the time horizon of climate change impacts may still be distant (although the physical impacts of climate change are increasingly crystalizing, and this has the potential to accelerate policy reactions and lead to the potential for transition risks). This can make it difficult to translate concerns regarding climate change into investment strategies or decisions and leads to reluctance to take action.

Second, investors ready to take climate change into account are confronted with a myriad of complexities. Even if progress has been rapid, current understandings of the effects of climate change are still limited, due both to data limitations and to the complexity of the issues at stake. In addition, investors face a lot of interrelated uncertainties (e.g., identifying the winning technologies when a number of them are still very much at an R&D stage, if not earlier, and in the context of a regime change in which large club effects/network-associated positive externalities are likely to arise;[24] assessing the sustainability and stability of the incentives provided by governments;[25] etc.).

Third, there has long been a lack of scalable solutions. Until now, most of the investment products addressing climate change were limited to asset classes such as private equity with narrow scalability opportunities for large institutional investors (pension funds, insurance companies, and sovereign wealth funds). In response to this, financial innovations have provided simple, scalable, pragmatic solutions that have proven readily implementable even though challenges are still pending. For example, these developments have enabled investors to take effective action in order to finance the low-carbon transition at a more relevant scale (green bonds) or manage transition risks in a rather rudimentary albeit robust manner (low-carbon indices).

As discussed above, green bonds provided a vector for scaling-up investments toward the transition. While offering the usual characteristics of a regular bond, green bonds enable investors to earmark funding for green projects, and, more generally, engage with issuers about their climate strategy. Further recent developments suggest that other green structures (covered bonds, securitization, and project bonds) could provide solutions to more directly channel capital toward green investments/assets. As an example, the International Finance Corporation (IFC) and Amundi have partnered to create a joint venture aimed at developing green finance in emerging markets. This innovative strategy will seek to foster both the supply and demand of green bonds in order to establish and develop green bond markets in emerging countries while promoting best practices.[26]

Well-designed low-carbon indices seek to track overall market returns while reducing climate change–related risks. Such indices help manage climate change risks over the long term without impacting returns in the short term by generating a (somewhat) free option on a mispriced asset (Andersson, Bolton, and Samama 2016a). The construction process is simple: the index screens corporations according to their climate change risk exposure and reduces the weighting of the most exposed companies while reinvesting the proceeds in the same sectors in order to retain sector and country neutrality. Hence, it enables investors to buy time at no additional cost, until the carbon-related risk becomes priced by markets.[27] Low-carbon indexes address all major climate risk challenges facing investors: time horizon (purchase of time for free), complexity (the indices focus on corporations that could be impacted instead of trying to identify the next successful green technology), and scalability (addressing a $10 trillion underlying market). First developed for and adopted by European asset owners (AP4, FRR)[28], low-carbon indexes have rapidly spread around the globe with, for example, their recent adoption by CalSTRS[29] and the New York Common Retirement Fund,[30] and GPIF[31]. This pragmatic approach recognizes the limited state of knowledge on matters such as the real impact of climate change on corporations while allowing investors to take their first steps in the direction of realigning their portfolios with a low-carbon economy.

The indices are an alternative to divestment strategies that may not be appropriate for institutional investors who need to maintain a market-wide exposure and/or have to recognize the uncertainties surrounding the understanding of climate change–related risks. They are also a complement to engagement as they enable investors to take reasoned action if and when engagement has not proven effective. Interestingly enough, from a public interest perspective, some indices, built on transparent rules, are generating an internal competition within each sector to accelerate the transition toward the low-carbon economy as those corporations that are excluded are incentivized by their reputation to do their best to rejoin the group of selected stocks.

5. Climbing the Learning Curve Together: Coalitions as Vectors of Knowledge Building and Sharing

Faced with big and complex issues that tend to overwhelm almost any individual institution and where collaboration may make sense (limited competition and/or aligned interest), financial institutions have formed coalitions. Such collective approaches have accelerated knowledge building and enabled knowledge sharing and dissemination.

While one can think of a number of institutions struggling with understanding the issues and with developing operational approaches to address them effectively, coalitions have been very effective in addressing challenges and testing innovative ideas.

For example, the PDC gathers investors who are seeking to implement decarbonization strategies and trying to align their portfolios with a low-carbon economy. The platform was intended to demonstrate the feasibility, diversity, and scalability of available solutions for managing the risks associated with climate change and bring this knowledge to the broader investor community. Within its first eighteen months, the initial $100 billion target was largely surpassed to reach $800 billion out of a total of $3.6 trillion in assets under management. This achievement and the 20 percent ratio (decarbonized portfolios/total assets) sent a clear signal that this approach was actually becoming mainstream. The PDC has organized various seminars[32] allowing its participants to review and discuss the variety of solutions (depending on constraints and objectives) to accelerate their learning curves by accessing each other's experiences.

6. Catalyzing Action: Innovation in Policy Making

The traditional role of governments with regard to climate change and the financial sector revolves respectively around designing and implementing (efficient, stable, and robust) climate policies to frame the low-carbon strategy and providing incentives/support to transition-related projects, on one side, and setting out regulations (complemented with supervision) on the other. Developing a more comprehensive financial sector policy agenda with regard to climate change involves taking an innovative approach to policy making, seeking to catalyze action that could complement climate policies while steering away from overburdening traditional financial regulation.

Indeed, as mobilizing financial markets is key to financing the energy transition, especially where public finance is scarce and there is a strong need for optimization,[33] governments are increasingly recognizing the power of private-sector financing. The first strong signal was sent at the COP21, highlighting the importance of climate-consistent finance: "making finance flows consistent with a pathway towards low greenhouse gas emissions and climate-resilient development" (Paris Agreement,

Article 2.1.c).[34] This was quickly followed by the green finance initiative under the G20's Chinese presidency (see GFSG 2016):[35] "we believe efforts could be made to provide clear strategic policy signals and frameworks, promote voluntary principles for green finance, (. . .) support the development of local green bond markets, promote international collaboration to facilitate cross-border investment in green bonds, encourage and facilitate knowledge sharing on environmental and financial risks, and improve the measurement of green finance activities and their impacts."

It can be said that policy makers have started to walk the talk with different initiatives implemented through stock exchanges (Singapore, London, etc.), supervisors (Brazil, the United Kingdom, Sweden, the Netherlands, France, etc.), or regulators (France, China).

Developing a Comprehensive Policy Agenda in an Advanced Economy Context: French and European Experiences

France developed a comprehensive climate strategy through its Energy Transition for Green Growth Act in 2015. The Act included provisions seeking to foster the appropriation of climate change–related issues by financial institutions by ensuring that relevant data (typically, from nonfinancial corporations) would become available for analysis and that banks and institutional investors (both asset owners and asset managers) would effectively step up their game with regard to these questions (Boissinot et al. 2016).

The provision for institutional investors has been trying to use disclosure as a nudge to (i) lead institutions to sketch out their strategic thinking and develop their analysis, and (ii) foster action through emulation and accelerated dissemination of best practices.

The regulation requires investors to disclose their climate (and more generally ESG) objectives, the analysis supporting the implementation of these objectives and to report on actions taken in that context.

France is an interesting example of how policy makers can leverage market forces to achieve goals that would previously have been approached in a more prescriptive way:

- It adopts a "comply-or-explain" approach, i.e., investors either have to provide information or explain the reasons why they chose not to do so.

- Recognizing that understandings of climate change–related issues are not mature enough to enshrine specific approaches, the French approach is not prescriptive with regard to methodology nor does it require any specific metrics to be reported on by financial institutions: entities can choose the methodology or metric they deem most relevant for their analysis of climate risks (and ESG criteria in general)—the only requirement is for institutional investors to justify their choice of approach and provide a clear description of their methodology and hypotheses.
- Public authorities suggest some criteria[36] but have chosen not to unilaterally impose any particular approach as a way of fostering still-needed innovations and accelerating the development of best practices in upcoming years.
- However, following a couple of reporting cycles, a government review is foreseen to assess the quality and effectiveness of the regulation, possibly leading to a revision of the secondary legislation. Although the "comply-or-explain" approach currently prevails, institutional investors face the prospect of a potentially more prescriptive approach if the current approach proves to be ineffective.

Ensuring that asset owners are effectively taking steps toward addressing relevant climate issues has a wide-ranging impact beyond prompting a demonstrated appropriation of the issues within the reporting institutions. On the one hand, it spreads the pressure over all asset classes equally (equity, fixed income, real estate, infrastructure, private equity, etc.). On the other hand, it also spills over globally as asset managers and other providers working with French asset owners may want to compete on providing methodologies to assess risks or investment solutions that, once developed for French clients, will be readily available everywhere. For instance, data providers are also incentivized to improve the quality and relevance of climate change–related data (which might not have existed before). In this regard, Article 173 creates impetus for higher-quality data and better analysis in order to meet investors' demands to publish their exposure to and strategy for addressing climate change–related risks.

The first assessments of Article 173 presented encouraging results and demonstrate the role and capacity of the French regulation to mobilize institutional investors while favoring a regional reach as its innovative aspects have already been acknowledged by the TCFD.[37]

Other initiatives included the first significant issuance of a sovereign green bond by the French debt management office in January 2017, a decision that was consistent with the wish to contribute to the definition of market standards and best practices from inside the market rather than through a regulatory approach that seemed inappropriate given the current maturity of the market.

Emulating the French initiative on climate change–related risk disclosure at the asset owner level could accelerate the shift within the financial sector and help to mainstream green finance. For instance, the generalization of the adoption of an Article 173–like resolution across Germany, the Netherlands, and the United Kingdom for example would impact over $10 trillion of assets,[38] and the extension to a greater circle including international states with large investor bases, such as Japan and China, would extend the impact of such a measure to a critical mass.

To that end, the European Union (EU) has taken a step toward climate change–related disclosure for asset owners within the directive, sketching out the regulatory framework for European pension funds (covering €2.5 trillion on behalf of around 75 million Europeans, around 20 percent of the workforce).[39] A provision was introduced in that context, setting out that pension funds must now, as part of their legal obligations, include climate risks and ESG risks in their decision-making processes. The directive has made reporting on ESG risks, and more specifically climate change–related risks and stranded assets, mandatory, and provides that the report shall be made public.[40] Member states must also ensure that pension funds provide information to prospective members on how ESG and climate factors are considered in their investment approaches. As the European Commission has sought advice from a dedicated high-level expert group, further policy developments are very likely (HLEG 2017).

Leveraging Green Finance for the Development of Emerging Economy Financial Markets: The Chinese Example

China's comprehensive approach to addressing climate change throughout its economy is remarkable.

China is conducting a fast-paced energy transition. Traditional climate policies set up in 2012 have since then progressively expanded carbon emission trading systems[41] laying the groundwork for a carbon tax to be imposed in 2020. Massive sums have been invested into upgrading renewable electricity capacity (China was the largest investor in renewable energies in 2016[42]), and thus it is safe to say China has been particularly active in its desire to mobilize capital markets (both at the national and international levels).

With regard to the mobilization of the financial sector, a People's Bank of China (PBOC)–led task force sketched out a number of options for developing a financial system that could support the Chinese transition while, in the meantime, more fundamentally upgrading the Chinese financial sector and ensuring that it can channel foreign money toward Chinese investment projects (see GFTF 2015). This agenda included a comprehensive roadmap: risk analysis, risk disclosure, green indices, green bonds, and was designed with the objective of enabling the private financing of the transition (the People's Bank of China clearly indicated that the public sector will not alone finance China's energy transition when communicating that 85 percent of the money needed to address environmental challenges would have to come from the private sector).[43] Interestingly, this initiative, designed to support domestic transition goals, has largely informed the G20's work on green finance.

The Chinese authorities were particularly innovative when it came to green bonds: China was the first country to publish the National Green Bond Guidelines in an attempt to ensure that a very significant part of the transition funding relies on financial markets (rather than public finance). This move was welcomed by market players and China has rapidly become the world's largest green bond market, with Chinese green bond issuances (including from China-based banks and corporations) accounting for 35 percent of total issuance volume for 2016.[44]

A growing number of developing countries are exploring the possibility of developing such a strategy that could become an integral part of a fully fledged NDC, contributing to deepening local financial markets as well as mobilizing local investors (public pension funds, public insurance companies, or sovereign wealth funds) in cooperation with multilateral development banks and foreign investors.

Recent developments have enabled the leveraging of investors to magnify and accelerate the global endeavor to address climate change.

Although ineffective by itself, this can very effectively complement climate policies, amplifying the policy signal in every corner of the economy. This movement has already started to shift the trillions needed for the reorientation of investment/reallocation of capital that underpins the global low-carbon transition. While asset owners already interested in climate change represent $100 trillion of assets under management,[45] each 0.1 percent shift from being interested toward taking action would mean a $100 billion reallocation of capital (equity or debt) consistent with a low-carbon economy.

Finally, climate change can be perceived as a prototype of a new and innovative policy approach that attempts to make the most of financial market forces through a more comprehensive understanding of interests that can be leveraged rather than the sole reliance on public resources and/or regulation. Current developments in green finance suggest that four pillars are critical to achieve this mobilization. Governments should seek to (i) frame the issue within a standard risk management approach, (ii) foster financial innovation, (iii) support peer pressure/transfer of knowledge, and (iv) play the role of catalysts. This four-pillar approach may find applications in other areas: the banking sector (see Bolton and Samama 2012), social inequalities, water, gender equality,[46] tobacco, etc. With the right approach, policy makers could mobilize vast amounts of money and align them with broader policy objectives, reinforcing price discovery mechanisms on longer-term topics, thereby perhaps leading to a new approach to capital markets by policy makers and contributing to a more sustainable form of capitalism.

Notes

The views and opinions expressed in this chapter are those of the authors and should not in any case be interpreted as reflecting those of any institutions they may be affiliated with. We thank Lauren Yeh, Timothée Jaulin, and Tobias Hessenberger for excellent research assistance. We express our gratitude to Patrick Bolton for the journey in mobilizing capital markets for public good that has inspired us over the years. We also wish to thank Mats Andersson, Pascal Blanqué, Jérôme Brouillet, Bob Buhr, Pascal Canfin, François Delattre, Philippe Desfossés, Laurent Fabius, Christiana Figueres, Carsten Frank, Huang Haizhou, Christopher Kaminker, Sean Kidney, Gildas Lamé, Philippe Le Houerou, Robert Litterman, Ma Jun, Jean-Marie Masse, Amélie de Montchalin, Xavier Musca, Adrian Orr, Janos Pasztor, Yves Perrier, Nicholas Pfaff, Fiona Reynolds, Nick Robins, Olivier Rousseau, Michael Sheren, Joseph Stiglitz, Sir Nicolas Stern, Christian Thimann, Laurence Tubiana, and participants at Portfolio Decarbonization Coalition seminars and retreats for comments and conversations.

1. At the time of this writing (June 2017).
2. See NASA Global Climate Change Database, http://climate.nasa.gov /evidence/.
3. Ibid.
4. See the survey (A.T. Kearney 2015) suggesting that American millennials aspire to be personally sustainable, and are willing to pay a premium for this, or the finding that millennials' desire to combine smart technology with sustainable lifestyles has already impacted current business models, and is set to continue. Sixty-one percent of millennials within the next five years want to sign up for a digital application allowing them to monitor their energy usage in their household (Dary 2016).
5. "Climate change is a global problem with grave implications: environmental, social, economic, political and for the distribution of goods [. . .]. It represents one of the principal challenges facing humanity in our day" ("Laudato si," Pope Francis, 2015).
6. "The challenges currently posed by climate change pale in significance compared with what might come. [. . .] So why isn't more being done to address it? [. . .] Climate change is the Tragedy of the Horizon. [. . .] In other words, once climate change becomes a defining issue for financial stability, it may already be too late" (Carney 2015).
7. "A sound ecological environment is the basic foundation for the sustainable development of humanity and society" (Xi 2014).
8. The two thresholds needed for the Paris Agreement to enter into force (over fifty-five countries representing at least 55 percent of the global GHG emissions) were reached on October 5, 2016. See http://unfccc.int/paris _agreement/items/9485.php.
9. For example, in 2016, the IEA (2016) presented evidence that renewables made up half of net electricity capacity addition in 2015 with costs expected to significantly drop in upcoming years; in the United States, 154 companies have signed the American Business Act on Climate Pledge to demonstrate their support for action on climate change and for the climate change Agreement in Paris. These companies employ nearly 11 million people, represent more than $4.2 trillion in annual revenue and have a combined market capitalization of over $7 trillion (as of January 1, 2017, see https://obamawhitehouse.archives.gov/the-press-office/2015/12/01 /white-house-announces-additional-commitments-american-business-act).

10. France issued its first "benchmark" green bond on January 25, 2017. This €7 billion 22-year green bond (the longest and largest ever) was a major success, being massively oversubscribed (demand was higher than €23 billion). Further tap issues (June and December) raised the total outstanding amount to €9.7 billion at the end of 2017. This issuance followed Poland's €750 million sovereign green bond in December 2016. See "IFR Offers a Triple Crown to France's Green OAT," December 9, 2017. http://www.aft.gouv.fr/articles/l-obligation-verte-de-la-france-triplement-recompensee-par-ifr_13117_lng2.html.

11. An IFC initiative is aiming at developing this market into emerging countries. See F. Samama, "Innovation in Climate Finance: World's Largest Green Bond Fund for Emerging Markets," " https://www.rockefellerfoundation.org/blog/innovation-climate-finance-launch-worlds-largest-green-bond-fund-emerging-markets/; see also "IFC, Amundi to Create World's Largest Green-Bond Fund Dedicated to Emerging Markets," https://ifcextapps.ifc.org/IFCExt/pressroom/IFCPressRoom.nsf/0/2CC3EDA1AE8B9B558525810900546887 and "IFC, Amundi Successfully Close World's Largest Green Bond Fund" at https://lnkd.in/gZgmwpY.

12. Eventually, as transition risks increase, one may imagine that the value of this signal rises and that a corporation being unable to issue a green bond, i.e., somehow failing to convince the market of its low-carbon strategy, will be discriminated against by lenders or debt investors.

13. See Paris Green Bonds Statement, December 9, 2015, available at: http://www.climatebonds.net/files/files/Paris_Investor_Statement_9Dec15.pdf.

14. Organized by CCLA Investment Management in 2012, the "Aiming for A" coalition is an investors' coalition of about fifty-five members, including the UK fund management industry and influential asset owners including the £100 billion a Local Authority Fund Management Forum and the largest members of the £12 billion Church Investors Group. The coalition's goal was to engage the ten major UK-listed utilities and extractive companies to earn an "A" in the CDP's Carbon Performance Leadership Index. The stated rationale for engaging asset owners was systemic risk management and a collective fiduciary duty to amplify long-term investors' voices and involve the ultimate beneficiaries of the company's sustainability behavior. Amundi joined the coalition and participated in the filing of the first adopted shareholder resolutions in 2015 at the General Assembly of BP and Shell.

15. BP's response, "Socially Responsible Investment," http://www.bp.com/en/global/corporate/investors/socially-responsible-investment.html, and Shell's is available at https://www.shell.com/investors/news-and-media-releases/investor-presentations.html.

16. See http://www.climateaction100.org/.

17. Launched on September 25, 2014, the Montreal Carbon Pledge is an investor commitment to measure and publicly disclose the carbon footprint of their investment portfolio on an annual basis. See http://montrealpledge.org/.
18. See https://www.churchofengland.org/media-centre/news/2016/05/%E2%80%98aiming-for-a%E2%80%99-climate-change-resolution-overwhelmingly-approved-by-glencore-shareholders.aspx/.
19. The PDC represented the entire finance sector at the COP21 Action Day plenary and announced that it has far exceeded its $100 billion decarbonization target and reached the $625 billion mark, see Amanda White, "Insitutional Investors Get Serious," December 9, 2015, https://www.top1000funds.com/news/2015/12/09/institutional-investors-get-serious-at-cop21.
20. "Risks resulting from climate change can be event driven (acute) or longer-term shifts (chronic) in climate patterns" (TCFD 2016).
21. "Transitioning to a lower-carbon economy may entail extensive policy, legal, technology, and market changes to address mitigation and adaptation requirements related to climate change. Depending on the nature, speed, and focus of these changes, transition risks may pose varying levels of financial and reputational risk to organizations" (TCFD 2016).
22. Less scrutiny has been devoted to physical risks although ongoing work is currently devoted to substantiate the impact of droughts, floods, rising sea levels, etc., in a number of countries.
23. The TCFD, composed of members chosen by the FSB, was set up "to develop recommendations for voluntary climate-related financial disclosures that are consistent, comparable, reliable, clear, and efficient, and provide decision-useful information to lenders, insurers, and investors." See the task force web site at https://www.fsb-tcfd.org/.
24. The poster child example of such risks is the development of electric vehicles (EVs) and other low-carbon mobility solutions: although it is clear that EV development is very likely and most probably a real business opportunity, differing models (e.g., in the respective roles of different solutions for addressing long-distance mobility) and competing standards makes it likely that some (currently well-positioned) players will not be able to realize the full potential of their products and strategy.
25. A number of early investors in renewables has felt strongly the changes or discontinuation of support schemes in some European countries when the euro crisis prompted fiscal consolidation, https://www.rockefellerfoundation.org/blog/innovation-climate-finance-launch-worlds-largest-green-bond-fund-emerging-markets/.
26. See "IFC, Amundi to Create World's Largest Green-Bond Fund," and Samama, "Innovation in Climate Finance."
27. And it must be noted that they have outperformed their benchmarks over the past few years.

28. "MSCI Launches Innovative Family of Low Carbon Indexes," September 16, 2014, https://www.msci.com/documents/10199/447d3ba7-e215 -45c9-8b14-74031a80f4bc.

29. See Ricardo Duran, "CalSTRS Commits $2.5 Billion to Low-Carbon Index," July 14, 2016, http://www.calstrs.com/news-release/calstrs-commits -25-billion-low-carbon-index.

30. See http://www.osc.state.ny.us/press/releases/dec15/120415.htm.

31. "Call for Applications for Global Environmental Stock Index," http://www.gpif.go.jp/en/topics/pdf/global_environmental_stock_index.pdf.

32. With the constant and valuable support of the Rockefeller Foundation.

33. For a discussion of the role of public finance in financing long-term investment, see Boissinot and Waysand (2012).

34. UNFCC, "Adoption of the Paris Agreement," https://unfccc.int/resource /docs/2015/cop21/eng/l09r01.pdf.

35. "G20 Leaders' Communique Hangzhou Summit," September 4–5, 2016, http://www.g20chn.org/English/Dynamic/201609/t20160906 _3396.html.

36. The secondary legislation provides a (nonexhaustive) typology of climate change–related risks. It also suggests reporting on contributions to the international/domestic climate objectives which, depending on the entities involved, could refer to analyzing the consistency of their portfolio with international targets and regional strategies and/or emphasizing their investments in assets that are considered to contribute to the low-carbon transition ("green assets").

37. Task Force on Climate-Related Financial Disclosures.

38. Based on Amundi Business Intelligence figures as of August 31, 2017.

39. The November 24, 2016 amendment to the 2003 Directive 2003/41/EC. The directive overhauls existing legislation, and aims to make the sector safer by introducing a new set of reporting and risk-management requirements. It also encourages long-term investments and seeks to facilitate fund ability to operate across European borders, removing legal barriers and lowering costs.

40. This assessment shall be made public through a publicly available reporting statement and must contain at least "the investment risk measurement methods, the risk-management processes implemented and the strategic asset allocation with respect to the nature and duration of pension liabilities, and how the investment policy takes environmental, social and governance factors into account."

41. In October 25, 2012, China launched the online platform for China Certified Emissions Reduction (CCER) trading, closely followed by the launch in 2013 of seven regional pilot emission trading schemes (ETSs)

that covered 9 percent of CO_2 emissions from energy use to provide a comprehensive reference for an integrated national ETS expected in 2017 as part of the 13th Five-Year Plan.

42. With a total investment of $286 billion across the world in renewable energy capacity, China's $102.9 billion investment representing 36 percent of the world total largely surpasses the United States ($44.1 billion) and the United Kingdom ($36.2 billion); see UNEP FI and Bloomberg, *Global Trends in Renewable Energy Investment 2016,* available at: http://fs-unep -centre.org/sites/default/files/publications/globaltrendsinrenewableen- ergyinvestment2016lowres_0.pdf.

43. See "Establishing China's Green Financial System—Background Paper A: A Theoretical Framework of Green Finance," People's Bank of China, United Nations Environment Programme, 2015.

44. Moody's Investors Service, January 18, 2017.

45. 827 investors representing about $100 trillion are backing CDP, see https://www.cdp.net/en/info/about-us.

46. See the experiments in Japan on Governance, Human Capital and Gender Equality topics with the setup of data and the creation of indexes integrating such factors and their use by large asset owners. See "GPIF Selected ESG Indices," July 3, 2017, http://www.gpif.go.jp/en/topics/pdf/20170703 _esg_selection_en.pdf.

References

Andersson, M., P. Bolton, and F. Samama. 2016a. "Hedging Climate Risk." *Financial Analysts Journal* 72 (3): 13–32

Andersson, M., P. Bolton, and F. Samama. 2016b. "Governance and Climate Change: A Success Story in Mobilizing Investor Support for Corporate Responses to Climate Change." *Journal of Applied Corporate Finance* 28 (2): 29–33

A. T. Kearney. 2015. "Demanding Millennials Buy Into Brands That Authentically Embody Their Values." *America@250 Series.*

Ban, K.-M. 2014. "Making Headway Against Climate Change." *Wall Street Journal* (September 25).

Boissinot, J., D. Huber, and G. Lame G. 2015. "Finance and Climate: The Transition to a Low-Carbon and Climate-Resilient Economy from a Financial Sector Perspective." *OECD Journal: Financial Market Trends* 1, 7–23.

Boissinot, J., D. Huber, I. Camilier-Cortial, and G. Lame G. 2016. "The Financial Sector Facing the Transition to a Low-Carbon Climate-Resilient Economy." *Trésor Eco* no. 185.

Boissinot, J., and C. Waysand, C. 2012. "Le financement des investissements de long terme: Quel rôle pour les pouvoirs publics?" *Revue d'économie financière* no. 108.

Bolton, P., and F. Samama. 2012. "Capital Access Bonds: Contingent Capital with an Option to Convert." *Economic Policy* 27 (70).

Caldecott, B., L. Kruitwagen, and D. MacDonald-Korth. 2016. *Summary of Proceedings: 4th Stranded Assets Forum at Waddesdon Manor.*

Calderon, F., and N. Stern. 2014. *Better Growth, Better Climate: The New Climate Economy Report.* Report of the Global Commission on the Economy and Climate.

Calderon, F., and N. Stern. 2015. *Seizing the Global Opportunity: Partnerships for Better Growth and a Better Climate.* Report of the Global Commission on the Economy and Climate.

Calderon, F., and N. Stern. 2016. *The Sustainable Infrastructure Imperative.* Report of the Global Commission on the Economy and Climate.

Carney, M. 2015. "Breaking the Tragedy of the Horizon—Climate Change and Financial Stability." Speech at Lloyd's of London (September 29).

CBI (Climate Bonds Initiative). 2016. *Climate Change: State of the Market 2016.* CBI Report.

CTI (Carbon Tracker Initiative). 2011. *Unburnable Carbon—Are the World's Financial Markets Carrying a Carbon Bubble?* CTI Report.

CTI. 2015. *Carbon Asset Risk: From Rhetoric to Action.* CTI Report.

Dary, Michael. 2016. "Millennials: Are You Ready for the New Faces of Energy Consumers?" The New Energy Consumer: Thriving in the Energy Ecosystem Series. Accenture Blog.

GFSG (Green Finance Study Group). 2016. *G20 Green Finance Synthesis Report.* Report to the G20.

GFTF (Green Finance Task Force). 2015, *Establishing China's Green Financial System.* GFTF Report.

Giraud, J., N. Lancesseur, and T. Roulleau, T. 2016. "Analyse Economique de l'Accord de Paris." *Trésor Eco* no. 187.

High-Level Expert Group on Sustainable Finance). 2017. *Financing a Sustainable European Economy (Interim Report).* Report to the European Commission.

Hong, H., F. W. Li, and J. Xu. 2016. "Climate Risks and Market Efficiency." NBER Working Paper, w22890.

ICMA (International Capital Market Association). 2017. *2017 ICMA Green Bond Principles.*

IEA (International Energy Agency). 2015. *Projected Costs of Generating Electricity.* IEA Report.

IEA. 2016. *Medium-Term Renewable Energy Market Report.* IEA Report.

IHS Markit. 2015. *Deflating the "Carbon Bubble."* IHS Markit Carbon Bubble Special Report.

IHS Markit. 2016. *Systemic Risk*. IHS Markit Strategic Report.

IMF (International Monetary Fund). 2013. *Energy Subsidy Reform: Lessons and Implications*. IMF Report.

IMF. 2015. *Counting the Cost of Energy Subsidies*. IMF Report.

Kaminker, C. 2016. *Analyzing Potential Bond Contributions in a Low-Carbon Transition*. OECD Report.

Nordhaus, W. D. 2016. "Projections and Uncertainties About Climate Change in an Era of Minimal Climate Policies." NBER Working Paper, w22933.

Pope Francis. 2015. *Laudato Si*. Encyclical Letter.

TCFD (Task Force on Climate-Related Financial Disclosure). 2016, *Recommendations of the Task Force on Climate-Related Financial Disclosure (Interim Report)*. Report to the FSB.

TCFD. 2017. *Recommendations of the Task Force on Climate-Related Financial Disclosure (Final Report)*. Report to the FSB.

Wolf, M. 2014. "A Climate Fix Would Ruin Investors." *Financial Times* (June 17).

UNEP and BNEF (United Nations Environment Program and Bloomberg New Energy Finance). 2016. *Global Trends in Renewable Energy Investment 2016*. UNEP and BNEF Report.

Xi, Jinping. 2014. *The Governance of China*. Beijing: Foreign Languages Press Co.

Contributors

Rabah Arezki is the chief economist of the World Bank's Middle East and North Africa Region and a Senior Fellow at Harvard's Kennedy School of Government. He was the Chief of the Commodities Unit in the IMF's Research Department. He has written on energy, commodities, development, and international economics. He has published widely in peer-reviewed journals and co-edited several books.

Karim El Aynaoui is managing director of the OCP Policy Center and advisor to the CEO and Chairman of the OCP, a global leader in the phosphate sector. He has published articles in scientific journals on macroeconomic issues in developing countries.

Nizar Baraka is president of the Economic Social and Environmental Council at the World Economic Forum and the Moroccan Minister of Economy and Finance.

Philippe Benoit is head of the Energy Efficiency and Environment Division at the International Energy Agency in Paris.

Jean Boissinot is director of Financial Stability at the Direction Générale du Trésor and associate professor of Economics at the Université Paris 1 Panthéon-Sorbonne.

Patrick Bolton is the Barbara and David Zalaznick Professor of Business and member of the Committee on Global Thought at Columbia University. He is the author of *Contract Theory* (MIT Press, 2005) with Mathias Dewatripont, the editor of *The Economics of Contracts* (Edward Elgar Publishing, 2008), and the co-editor of *Credit Markets for the Poor* (Russell Sage Foundation, 2005) with Howard Rosenthal.

Ujjayant Chakravorty is professor of Economics at Tufts University and Fellow at the Toulouse School of Economics and CESifo. He is co-editor of the *Journal of Environmental Economics and Management*. He is co-editor of *India and Global Climate Change* (Oxford University Press, 2003). His work has also appeared in the *American Economic Review*, *Journal of Political Economy*, *Econometrica*, *Journal of Environmental Economics and Management*, and the *Journal of Economic Dynamics and Control*.

Thierry Déau is chairman and CEO of Meridiam.

Christian Gollier is a professor at the Toulouse School of Economics and a visiting professor at Columbia University. He is the author of *Ethical Asset Valuation and the Good Society* (Columbia University Press, 2017), *Pricing the Planet's Future: The Economics of Discounting in an Uncertain World* (Princeton University Press, 2012), and *The Economics of Risk and Time* (MIT Press, 2004). He is also a lead author of the fourth and fifth reports of the Intergovernmental Panel on Climate Change.

Bård Harstad is professor of economics at the University of Oslo. His work has appeared in the *American Economic Review*, *Quarterly Journal of Economics*, *Journal of Political Economy*, *Review of Economic Studies*, and *American Political Science Review*, among others.

Ted Loch-Temzelides is professor of economics and a Baker Institute Center for Energy Studies Rice Scholar at Rice University. His work has appeared in *Energy Economics*, *Quantitative Economics*, *The Energy Journal*, *Economics of Energy and Environmental Policy*, and elsewhere.

Maurice Obstfeld is the economic counsellor and director of research at the International Monetary Fund, and the Class of 1958 Professor of Economics at the University of California, Berkeley. He is the co-author of *International Economics* (Pearson, 2014) with Paul Krugman and Marc Melitz, and of *Foundations of International Macroeconomics* (MIT Press, 1996) with Kenneth Rogoff.

Ian Parry is the principal environmental fiscal policy expert in the Fiscal Affairs Department of the International Monetary Fund. He is the co-editor of *Fiscal Policy to Mitigate Climate Change: A Guide for Policymakers* (IMF, 2012) and the co-author of *Issues of the Day: 100 Commentaries on Environmental, Energy, Transportation, and Public Health Policy* (Routledge, 2010) with Felicia Day.

Frédéric Samama is Co-Head of Institutional Clients Coverage at Amundi and Founder of the Sovereign Wealth Fund Research Initiative.

Katheline Schubert is professor of economics at the University Paris 1 Panthéon-Sorbonne and associate chair at the Paris School of Economics. Her work has appeared in the *American Journal of Agricultural Economics, Energy Journal, Journal of Economic Theory, Journal of Mathematical Economics, Macroeconomic Dynamics,* and elsewhere.

Julien Touati is a partner at Meridiam.

Rick van der Ploeg is professor of economics and research director of the Oxford Centre for the Analysis of Resource Rich Economics at the University of Oxford. He is the co-editor of *The Economics of Resource Rich Economies* (Edward Elgar, 2015) with A. J. Venables, and *Climate Policy and Non-Renewable Resources—The Green Paradox and Beyond* (MIT Press, 2014) with K. Pittel and C. Withagen.

Martin L. Weitzman is professor of economics at Harvard University. He is the author of *The Share Economy* (Harvard University Press, 1984), *Income, Wealth, and the Maximum Principle* (Harvard University Press, 2003), and *Climate Shock* (Princeton University Press, 2015) with Gernot Wagner.

Index

Page numbers in italics indicate tables or figures.